THE RAND
CORPORATION

HARVARD POLITICAL STUDIES

Published under the direction of the
Department of Government in Harvard University

THE RAND
CORPORATION

Case Study of a Nonprofit Advisory Corporation

Bruce L. R. Smith

HARVARD UNIVERSITY PRESS
CAMBRIDGE·MASSACHUSETTS·1966

TO MY MOTHER AND FATHER

PREFACE

The idea of undertaking this study first occurred to me in the spring of 1961. It struck me that the rise of the novel RAND-type advisory corporation was an interesting phenomenon that had never been seriously considered in the scholarly literature. What was the import of this development for the American governing system? Was it merely a temporary aberration from the "normal" pattern of government organization? Or, with the growing impact of science and technology on the structure of government and the process of policy formation, was increasing reliance on novel "RAND-type" organizational forms destined to become a prominent feature of modern government? What light did the experience of RAND and similar organizations throw on the twentieth-century problem of relating the expert to the politician within some framework of public accountability? These questions seemed to me of absorbing interest, and I thought they would provide a challenging focus for doctoral dissertation work. After completing my general exams in May 1961, I went to Don K. Price, Dean of the Graduate School of Public Administration and Professor of Government at Harvard University, to discuss with him the possibility of doing a study of the nonprofit advisory corporations. Dean Price, whose pioneering work on government-science relations had done so much to focus attention on this new class of problems, kindly agreed to act as my faculty advisor for the study.

I began the research while serving as a management trainee in the Office of the Secretary of Defense during the summer of

1961. The initial phases of the research consisted mainly in interviews with numerous Defense officials and people at various research organizations in the Washington area. The interviews granted me by various individuals in and out of the government were of great assistance in providing background and a frame of reference for the study. (In all, in the period 1961-1963, I conducted over 400 interviews with persons from the government, nonprofit research institutions, industrial and other profit-making research institutions, and the academic community.) Special thanks are due to Burton B. Moyer, Jr., Training Officer of the Office of the Secretary of Defense, who permitted me occasional time out from other duties to pursue my research interests and who assisted me in numerous other ways in carrying out parts of the study. I also wish to thank my fellow management trainees who gave me the benefit of their views in many a spirited bull session. My friend C. Payne Lucas, now with the Peace Corps in Africa, took the trouble to prepare a detailed topical outline that was very helpful.

My research was facilitated throughout by the fact that, as a Defense Department employee, I had a clearance and was able to move easily through the classified community. However, it should be made clear that this study is based exclusively on unclassified sources and materials.

In the winter of 1961, under a grant from the Science and Public Policy Seminar of Harvard University, I made a three-week trip to RAND headquarters in Santa Monica, California. During that trip, Dr. Alexander L. George, then head of RAND's Social Science Department, invited me to come to Santa Monica the following summer as a temporary graduate student consultant to work on a project dealing with the dynamics of political development in the transitional societies. I was able to spend some portion of my time during the summer pursuing my own research interests as well as observing RAND operations first-hand.

The trustees and management of the RAND Corporation cooperated to an unusual degree in making this study possible. I was granted wide access to internal RAND materials, the opportunity to observe many phases of RAND's internal operations, and full freedom to interview people throughout all levels of the organization. My study was, however, in no sense an "official" or "sponsored" study. There was never anything in the nature of an official management surveillance or review of my work. Needless to say, I never received any financial assistance from RAND or any other advisory organization (except for my brief consulting stint in the summer of 1962) in connection with study's preparation. My understanding with the RAND Corporation was that the study was entirely an independent undertaking done as my doctoral dissertation in the Department of Government at Harvard University. RAND management admirably lived up to their agreement, although I dare say that they must have been sorely tried at times by my presence and endless questions, requests for data, and general intrusions into the life of the organization. I am afraid that I trespassed on the time and patience of many RAND people far beyond the limits of courtesy.

I also benefited greatly from the cooperation of the management and staff of other nonprofit advisory corporations. During the period 1961-1963, I visited the Stanford Research Institute, Analytic Services, Inc. (ANSER), the System Development Corporation, the Hudson Institute, the Center for Naval Analysis (and the old Operations Evaluation Group), the Research Analysis Corporation, the Institute for Defense Analyses, Planning Research Corporation, the Institute for Strategic Studies in London, the Weapons Systems Evaluation Group, and received friendly assistance from numerous research and management personnel associated with these organizations.

It is impossible to acknowledge by name all the people who helped me at various stages in the study's preparation. I am sure that they will recognize the extent of my indebtedness and will not feel slighted by a general expression of gratitude for the fine cooperation which made the research for this study more a pleasure than a chore. A few individuals, however, must be mentioned: Edward L. Bowles, formerly special consultant to General H. H. Arnold and one of Project RAND's original founders, now of the Massachusetts Institute of Technology and the Raytheon Company; L. J. Henderson, R. D. Specht, David Novick, J. R. Goldstein, Elizabeth Oswald, the late John D. Williams, Joseph M. Goldsen, Alexander L. George, Bernard Brodie, Burton H. Klein, Amrom Katz, George Rosen, Charles A. H. Thomson, George Clement, John C. Hogan, and Brownlee W. Haydon of the RAND Corporation; Arthur E. Raymond of the Douglas Aircraft Co.; Albert Wohlstetter, formerly of the RAND Corporation and now University Professor in Political Science at the University of Chicago, and Roberta Wohlstetter; Robert A. Levine, formerly of the RAND Corporation and now with the Office of Economic Opportunity; Stephen Enke and James E. King of the Institute for Defense Analyses; Herman Kahn of the Hudson Institute; Stanley Lawill of ANSER; Charles Mottley, Leon Sloss, and the late W. D. Eisner of the Stanford Research Institute; Charles Bray of the Smithsonian Institution; Alastair Buchan of the Institute for Strategic Studies in London; Samuel Clements and Edward L. Katzenbach of the Department of Defense; Lynn E. Baker of the Department of the Army, Research and Development; Henry S. Rowen and Edward K. Hamilton of the Bureau of the Budget; George Pettee of the Research Analysis Corporation; Hascal Wilson,

Thomas Phipps, and Thomas Milburn of the Naval Ordnance Test Station; Frank Bothwell, Jacinto Steinhardt, and William Meckling of the Center for Naval Analysis (and OEG); Col. Wesley W. Posvar, Col. Marshall E. Baker, Lt. Col. John G. Dailey of the U.S. Air Force; W. Phillips Davison of the Council on Foreign Relations; Lt. Gen. Donald Putt (USAF, ret.) of United Technology; and Thomas C. Schelling, Morton H. Halperin, John M. Gaus, W. Barton Leach, and William Y. Elliott of Harvard University.

Above all, I owe an enormous debt to Don K. Price. Dean Price's kindly, patient, yet exacting, guidance was invaluable to me. He helped me to make the connections which I hadn't the imagination to see. He stimulated me to press further when I sought to take refuge in equivocation. Dean Price's lucid mind, broad learning, and deep human concern influenced my development more than he knows.

Robert G. McCloskey, then chairman of the Government Department at Harvard University, encouraged me in my work and arranged for a grant from the Frank G. Thomson Bequest which enabled me to do some additional field research in the summer of 1963. Part of my research was also supported by the Samuel Fels Fund, and I am grateful to Mr. Dale Phalen, Executive Secretary of the Fels Foundation, for his interest and support.

In June 1964, after completing the Ph.D., I joined the RAND staff as a Research Associate in RAND's Social Science Department. I have since been engaged in research on the effects of technology on government organization, management of R & D activities by the Air Force Systems Command, various aspects of science planning, and problems of organizational innovation. In early 1965 I took a three-month leave of absence from RAND, under a grant from the Science and Public Policy Seminar of Harvard University, to revise my thesis for book publication. I did not attempt to update the book substantially or to introduce new material or additional data derived from personal experiences at RAND. There have been some interesting recent developments at RAND, but I thought that to deal adequately with these would be beyond the scope of the present study. Moreover, in all important respects—on the interesting questions of relationship to client, problems of internal management, the role of analysis in policy formation—I believe the treatment presented here still stands and represents my current views.

Daniel Bell of Columbia University read the last chapter of the revised manuscript and offered useful comments. My wife, Elise, deserves thanks for putting up with a rather trying husband during the final stages of the writing. In addition, she read parts of the manuscript and volunteered a number of helpful editorial suggestions. Finally, I must add the conventional acknowledgment—not likely to be seriously disputed—that I alone am responsible for whatever errors of fact or conception that remain in the book.

Pacific Palisades, California   Bruce L. R. Smith
November 1965

CONTENTS

THE RAND
CORPORATION

Case Study of a
Nonprofit Advisory Corporation

ABBREVIATIONS

AEC	Atomic Energy Commission
AID	Agency for International Development
ANSER	Analytic Services, Inc.
ARPA	Advanced Research Projects Agency
ASWORG	Antisubmarine Warfare Operations Group
CEIR	Council for Economic and Industrial Research
CNA	Center for Naval Analysis
DASA	Defense Atomic Support Agency
HumRRO	Human Relations Research Office
IDA	Institute for Defense Analyses
ISA	International Security Affairs (OSD)
LMI	Logistics Management Institute
NASA	National Aeronautics and Space Administration
OEG	Operations Evaluation Group
ORO	Operations Research Office
OSD	Office of the Secretary of Defense
OSRD	Office of Scientific Research and Development
RAC	Research Analysis Corporation
SAC	Strategic Air Command
SDC	System Development Corporation
SOFS	Strategic Offensive Forces Study
SORO	Special Operations Research Organization
SRI	Stanford Research Institute
WSED	Weapons System Evaluation Division
WSEG	Weapons System Evaluation Group

► I ◄

INTRODUCTION

If our future depends upon the effectiveness of our
research effort, then it would seem to follow that one
of the first things we need to research is the effectiveness of different styles of research organization and
operation.

—*Dwight Waldo, Panel Comment, Brookings Institution*
Dedication Lectures, 1961

The emergence of the nonprofit research or advisory corporation is one of the most striking phenomena of America's
postwar defense organization. A small but influential community, made up largely of civilian researchers and strategists, now provides defense policy makers with advice on a
wide range of important problems. Leading members of this
community have come to play a role in defense policy formation scarcely imaginable several decades ago. Struck by the
phenomenon, one foreign observer remarked that representatives of certain U.S. research organizations roam through
the corridors of the Pentagon "rather as the Jesuits through
the courts of Madrid and Vienna three centuries ago."[1]
Each of the three military services has several "nonprofits"
performing analytic and advisory services on a contract basis.
The Navy has the Center for Naval Analysis, which in-

[1] Quoted in Daniel Bell, ed., *The Radical Right*, rev. ed. (Doubleday
Anchor, Garden City, N.Y., 1964), p. 33.

1

cludes several component organizations, administered on contract by the Franklin Institute of Philadelphia. The Center for Naval Analysis grew out of the old Operations Evaluation Group (OEG), which had its origins in 1942 as the Antisubmarine Warfare Operations Group (ASWORG) set up to help combat the submarine threat to Allied shipping in the Atlantic.[2]

The Army has sponsored the creation of the Research Analysis Corporation (RAC), Human Relations Research Office (HumRRO), and Special Operations Research Organization (SORO). In addition, the Army has a contract-research group furnished by the Stanford Research Institute (SRI) and a contract-research operation also in effect with Technical Operations, Inc. The Army came rather tardily to recognize that civilian expertise could enhance its fighting capacity, and has been a comparative latecomer in the use of nonprofit corporations for technical advice and research assistance. Unlike her sister services, the Army had no effective program of operations research during World War II either at the Washington staff level or in the field.[3] In 1948, the Army first established an advisory organization: the Operations Research Office (ORO). The organization was

[2] James Phinney Baxter, *Scientists Against Time* (Little, Brown, Boston, 1947); Jacinto Steinhardt, "The Role of Operations Research in the Navy," *United States Naval Institute Proceedings* (May 1946); Florence Trefethen, *The History of Operations Research Research* (Operations Research Office, Chevy Chase, Md., 1953); and Samuel Eliot Morison, *The Two-Ocean War* (Little, Brown, Boston, 1963), pp. 102-109.

[3] The U.S. Army did make a token effort to apply operations research techniques in the Pacific theater near the end of the war, but this effort did not amount to much. See Colonel Seymour I. Gilman, USA, "Operations Research in the Army," *Military Review* 36:54-64 (July 1956). In Britain, operations research also was latest in coming to the Army, although it appears that the British Army did make somewhat more effective use of operations research in World War II than the U.S. Army. See Ronald Clark, *The Rise of the Boffins* (Phoenix House, London, 1962).

administered through a contract with the Johns Hopkins University, although in practice the ORO functioned virtually as an autonomous unit. The arrangement with the university lasted until the summer of 1961 when the Army severed the contract and reestablished the ORO as a new nonprofit research corporation: the Research Analysis Corporation (RAC).[4]

The Air Force is perhaps the biggest user of nonprofit corporations for advisory services and other technical assistance among the military services. This is not entirely surprising. For the Air Force as the youngest of the services found itself without the historic technical bureaus that had grown up in the Army and Navy, and perforce had to rely on extensive contract assistance for many services in its formative stages. The fact that the Air Force had no traditional pattern of internal organization also meant that the Air Force was less tied to precedent and could experiment more easily with novel organizational forms.

The Air Force's principal advisory organization is the RAND Corporation, perhaps the best known of the defense organizations. RAND emerged at the end of World War II when it was provisionally organized as an adjunct of the Douglas Aircraft Company in Santa Monica, California. In 1948, RAND separated from Douglas Aircraft and was incorporated under the laws of California as a private nonprofit corporation.

Another Air Force advisory corporation, Analytic Services, Inc. (ANSER) of Arlington, Virginia, was established under RAND sponsorship in 1958. ANSER is much smaller in size and much more limited in the scope of its operations than RAND, and has enjoyed less freedom to perform work of

[4] *New York Times*, May 28, 1961, p. 1, col. 4. The new contract with RAC became effective on September 1, 1961.

its own choosing than has RAND throughout most of its history. The Air Force also contracts with a number of other nonprofit corporations but for various technical services that differ rather sharply from the research missions of RAND and ANSER. The System Development Corporation (SDC), for example, provides a technical-support function for the Air Force in particular aspects of information-systems design, especially with respect to the Air Force's air defense mission. Nonprofit corporations like Aerospace and MITRE perform yet another function, namely, managerial assistance to particular Air Force commands in the technical direction and supervision of the work of other contractors.

At the level of the Office of the Secretary of Defense, several organizations have developed to parallel the service-affiliated advisory institutions. The first of these—the Weapons Systems Evaluation Group (WSEG)—had its origins in the early postunification controversies over the roles and missions of the respective services. At the Key West and Newport Conferences convened in 1948 to resolve the roles and missions dispute, it was decided to establish a group which could objectively assess the claims for competing weapons systems advanced by the different services.[5] WSEG was established within the government under the now defunct Research and Development Board, but worked under the direction of the Joint Chiefs of Staff. Unfortunately, for a variety of reasons WSEG got off to a poor start and in its early years never was able to function effectively.

The Rockefeller Report on Defense Reorganization of 1953

---

[5] Timothy W. Stanley, *American Defense and National Security* (Public Affairs Press, Washington, 1956), p. 89. See also George E. Pugh, "Operations Research for the Secretary of Defense and the Joint Chiefs of Staff," *Journal of the Operations Research Society of America* 8:839-840 (November-December 1960), and Walter Millis, ed., *The Forrestal Diaries* (Viking, New York, 1951), p. 477.

called attention to several weaknesses in the WSEG operation, and suggested some important changes in the group's staffing and functioning.[6] Then in 1955 a Subcommittee Report on Department of Defense Research Activities of the Second Hoover Commission urged that "the only way a staff of sufficient size and competence for WSEG can be assembled is by a contract operation with a university or nonprofit organization."[7] Later in the year, the Department of Defense contracted with the Massachusetts Institute of Technology to provide technical personnel to support the WSEG operation. In 1956, the Institute for Defense Analyses (IDA) was formed to act as technical backstop to WSEG and to facilitate the recruitment of high-caliber scientific manpower. IDA is a loose holding company of universities (initially five but since expanded to eight) incorporated under the laws of Delaware as a nonprofit corporation. Although initially established as a supporting agency for WSEG, it has broadened its functions to a point where IDA's supporting division to WSEG was only one of five IDA working divisions. In 1962 IDA was the principal advisory organization serving the Office of the Secretary of Defense as a whole. WSEG, supported by the Weapons System Evaluation Division of IDA, acted as a research arm for the Joint Chiefs of Staff and for the Director of Defense Research and Engineering. The Logistics Management Institute (LMI), created in 1962, served as a research aid to the Assistant Secretary of Defense for Installations and Logistics.

In addition to the above organizations, working more or less on a permanent basis for a defense agency or agencies,

---

[6] *Report of the Rockefeller Committee on Department of Defense Organization*, Senate Committee on Armed Services, 83rd Cong. 1st Sess., April 11, 1953.

[7] *Subcommittee Report on Research Activities in the Department of Defense and Defense Related Agencies*, prepared for the Commission on Organization of the Executive Branch of the Government, April 1953, p. 30.

a large number of other nonprofit organizations perform various *ad hoc* studies on contract for defense agencies. The major defense agencies commonly have one or more permanent contract advisory organizations, and from time to time contract for special advisory services with other organizations and individual consultants. As the business of providing analytical advice to defense policy makers has flourished since World War II, the number of organizations engaged in such work (or seeking to enter the field) has increased substantially.[8]

OPERATIONS RESEARCH AND SYSTEMS ANALYSIS

The character of the work performed by the principal defense advisory organizations usually has centered around "operations research"—or some variation or outgrowth of the analytical techniques associated with that concept. Operations research was developed during World War II (though something similar to it had been used in industrial management for some years) to provide quantitative aids to defense decision makers. The assistance provided by operations research aimed first at optimizing the operational employment of existing weapons (or other military) systems, and later broadened into more general tasks of rationally allocating resources for national security purposes.

Studies of Allied anti-submarine tactics and the proper placement and use of radar crews in the initial phases of

---

[8] Edward L. Katzenbach, "Ideas: A New Defense Industry," *The Reporter*, March 2, 1961, pp. 17-21. There are no reliable estimates of the exact numbers of such organizations currently in existence. Claude Witze (in *Air Force Magazine*, November 1961, p. 54) has asserted that "the demand has resulted in the organization of more than 350 nonprofit organizations in the postwar years, most of them concerned with some aspect of the missile, electronic, and atomic technologies." But this figure has a spurious exactness. There are no data which substantiate this estimate, and idiosyncracies of definition could in any case substantially expand or contract the number of "nonprofit organizations."

the war were among the most famous early applications. It was found, for instance, through a statistical analysis of depth-charge "kills" that altering the depth at which the charge was exploded would greatly enhance the probability of destroying the enemy submarine. Similarly, operations-research studies played an important role in overcoming the obstacles to effective employment of the first British radar sets in 1940. Difficulties were encountered initially in co-ordinating the use of radar with anti-aircraft batteries and with the operations of interceptor aircraft. As the war progressed, operations-research techniques found many significant new uses and played no small role in the ultimate success of the Allied war effort.[9]

[9] For the early history and general nature of operations research, see: Philip M. Morse and George E. Kimball, *Methods of Operations Research* (John Wiley, New York, 1951); Sir Robert Watson-Watt, *The Pulse of Radar* (Dial, New York, 1959), and *Three Steps to Victory* (Odhams Press, London, 1957); Derek Wood and Derek Dempter, *The Narrow Margin* (McGraw-Hill, New York, 1961); J. G. Crowther and R. Whiddington, *Science at War* (HMSO, London, 1947); P. M. S. Blackett, "Operational Research," *The Advancement of Science* 5:26-38 (April 1948); Charles J. Hitch, "Sub-Optimization in Operations Problems," *Journal of the Operations Research Society of America* 1:87-99 (May 1953); and the works cited in note 2 above. Those wishing a more complete literature on the subject are referred to Vera Riley, *An Annotated Bibliography on Operations Research* (ORO, Chevy Chase, Md., June 15, 1953); Harvard Defense Policy Serial No. 106, "Fact-Finding and Analytical Facilities Available to Policy-Makers in the Defense Establishment," compiled by W. Barton Leach, et al. (mimeo); and various issues of the *Journal of the Operations Research Society of America*, published since 1950, most of which contain bibliographical material.

Some trace the origins of operation-research techniques applied to military affairs back to works published as early as 1882 and 1916 but little heeded then or for many years. See F. W. Lanchester, *Aircraft in Warfare: The Dawn of the Fourth Army* (Constable, London, 1916), and W. R. Livermore, *The American Kriegsspiel: A Game for Practising the Art of War Upon a Topographical Map* (Houghton, Mifflin, Boston, 1882).

Good expositions of the intellectual antecedents to systems analysis are Roland N. McKean, *Efficiency in Government Through Systems Analysis: With Emphasis on Water Resource Development* (John Wiley, New York, 1958), especially pp. 5-21 and Herbert A. Simon, *The New Science of Management Decision* (Harper, New York, 1960).

Despite a deep reservoir of scientific talent, the Germans apparently had no counterpart to the Allied operations researchers. Nazi decision making, reflecting the Führer's personal style, tended to be more intuitive and irrational. This cost the Germans heavily on occasion. A crucial delay in the German missile-development program, for example, resulted from a Hitler dream that no V-2 would ever reach England.[10] Florence Trefethen observes:

> The share which the total operations research effort had in bringing final victory to the Allied cause cannot be measured. It was, however, an asset peculiar to the Allies and seems to have had no rival or counterpart within the organizations of the enemy forces. It represented, in effect, a view of war which was antithetical to Hitler's, bringing measurement, control and the mathematical analysis of complex operations into play against the romantic and "inspired" moves of the axis forces.[11]

Since World War II, as military planning has become more intricate, the character of useful advisory work has undergone a transformation. A broader and more refined sister discipline—"systems analysis"—has evolved out of the earlier operations-research techniques. The systems-analysis approach differs significantly from traditional World War II operations research in being less quantitative in method and more oriented toward the analysis of broad strategic and policy questions, and particularly in seeking to clarify choice under conditions of great uncertainty.[12] Traditional opera-

---

[10] Walter Dornberger, *V-2* (Ballantine Books, New York, 1954), translated by James Cleugh and Geoffrey Halliday, p. 87.

[11] Quoted in Harvard Defense Policy Serial No. 106, Part D, p. 2. Bernard and Fawn Brodie comment: "His contempt for the expert and his reliance upon his own 'creative genius' were in the end to prove fatal for Hitler and for his kind of Germany." *From Crossbow to H-Bomb* (Dell Books, New York, 1962), p. 200. See also Clark, *The Rise of the Boffins*, for a detailed discussion of how operations research aided the Western Allies against the Germans.

[12] The differences between traditional operations research and the broader

tions research tends toward elaborate mathematical techniques and the analysis of low-level tactical problems where objectives are precisely stated and admit of simple solution. A military problem amenable to treatment by operations research can be "solved," that is, a precise solution can be found that will represent an optimum allocation of resources at the decision maker's disposal. In contrast, systems analysis aims at a range of problems to which there can be no "solution" in a strict sense because there are no clearly defined objectives that can be optimized or maximized. The systems-analysis approach retains, however, elements of rigor and preciseness that offer more reliable assessments of numerous difficult choices than simple intuition or inference from unexamined precept. It is evident that few problems at the higher policy echelons are amenable to fruitful treatment by simple quantitative techniques alone.

In practice, many of the advisory corporations do analytical work of both kinds, since certain kinds of questions can be treated by precise optimizing techniques while another range of questions must be dealt with by the broader approach. Also, the broad systems analysis may contain subelements within the larger analytical context where operations-research techniques are useful in providing specific

systems-analysis approach are ably expounded in James R. Schlesinger, "Quantitative Analysis and National Security," *World Politics* 15:295-315 (January 1963). Albert Wohlstetter sums up the differences between the two approaches by attributing to systems analysis (1) a more distant future environment, with greater flexibility for choice, (2) more interdependent variables, (3) greater uncertainties, and (4) less obvious objectives and rules of choice. "Scientists, Seers, and Strategy," unpublished monograph (Council for Atomic Age Studies, Columbia University, 1962), pp. 36-37. A condensed version of this study appears as "Strategy and the Natural Scientists," in Robert Gilpin and Christopher Wright, eds., *Scientists and National Policy-Making* (Columbia University Press, New York, 1964), pp. 174-239. For a discussion of the meaning of "system" in systems analysis, see Chapter V, pp. 160-162.

answers to narrow questions. Thus, although systems analysis normally deals with a range of problems to which there are no unambiguous solutions, specialized research techniques can sometimes clarify important aspects of a broad problem and reduce the uncertainty confronting the analyst and the policy maker. The task of systems analysis work, indeed, is principally that of integrating a variety of specialized research skills into a common analytical framework bearing on an important problem. The aim of such work is to be precise where possible and, where precision is ruled out by a context of conflicting values and unclear objectives, analysis strives to identify the many complex dimensions of a critical policy choice and to help the policy maker assess the costs and consequences of alternative courses of action.

Although one should recognize the interrelationship between the quantitative and the qualitative aspects of useful analytic advice, it is important to keep clearly in mind a conceptual distinction between the two. Much confusion has resulted from a failure to do so. It is sometimes supposed, for example, that various advisory corporations attempt to analyze all national security problems in precise quantitative terms. This view rests on the misconception that analytic "solutions" can be obtained to broad problems where there are no clearly defined ends or objectives to be sought. Unless values or objectives are "given," no optimal solutions are possible. And the political process by definition is concerned with value conflicts: the "political" is the process of defining (and redefining) community values and searching for appropriate ends to pursue amid a welter of conflicting and unclear objectives. It follows that no amount of research or analysis can provide the policy maker with simple guides to action on the muddy normative issues of high policy. At best analysis usually can only clarify the bases for decision, discipline the policy maker's intuition, and enrich and deepen his understanding of a complex problem.

At the same time it is easy to err in the opposite direction and to undervalue the role of analysis. Witness the recent pronouncement by a leading student of defense affairs: "Good military policy is *only* the product of brave choice and ingenious compromise by experienced politicians."[13] Beneath the rhetoric of this statement lies a dangerous and untenable assumption: that in an age of explosive advances in technology the United States can continue to rely largely on conventional wisdom, civilian or military, in making critical national security decisions. Policy choices in this area have become so complex, and so crucially influenced by developments at the frontiers of science and technology, that no sensible policy maker can operate without extensive research and analytical aids.

A rapidly evolving technology carries with it far-reaching policy implications which the policy maker, however judicious and experienced, can hardly begin to assess without analytical assistance. Forward-looking strategic thinking is urgently necessary as the political environment changes dramatically in response to developments in military technology. Experience counts for relatively less because often there is no experience comparable to what the policy maker must face. And with weapons of mass destruction in national arsenals there is no longer the luxury of allowing for a margin of error in critical aspects of the nation's defense policies. Efforts to reduce the uncertainty confronting the policy maker therefore can only be welcomed. Moreover, in contrast to the situation in World War II when crucial decisions had to be made under enormous time pressures, the cold war era has

---

[13] Samuel P. Huntington, *The Common Defense* (Columbia University Press, New York, 1961), p. xii. The italics are the author's, not mine.

frequently afforded an opportunity to carry out complex studies over a period of time and to have the results aired at numerous policy levels.[14] Parameters of a difficult problem assumed to be unknowable may turn out to be knowable, or at least more comprehensible, if time is available for data-gathering and analysis. The result is that official thinking can be ventilated at a number of levels by a healthy infusion of outside ideas. The policy makers are provided with useful new ways of looking at complex problems. The analytic product, although seldom a clear guide to decision, serves nevertheless as a valuable aid to sophisticated judgment. As policy choices become ever more complex, the intellectual "backup" to policy action can only be expected to increase. Indeed, Daniel Bell has pointed to a growing penetration of research and intellectual activity into all aspects of social life as the distinctive characteristic of the "post-industrial society" into which the United States seems to be moving.[15]

Corresponding to the evolution of the systems-analysis approach out of earlier operations-research techniques, there have been noticeable changes in the professional staffs of various advisory corporations. Whereas in World War II mathematicians and physical scientists were the dominant skill groups, economists have increasingly gained in numbers and importance, and have come to play a leading role. As the concern of many advisory corporations shifts more toward "limited" war problems, it is possible that the more qualitative social scientists will witness a similar growth in numbers and influence. The economists could act as the

14 For an elaboration of this point, see Wohlstetter, "Strategy and the Natural Scientists," pp. 181-183. See also Chapter VI.

15 Daniel Bell, "The Post-Industrial Society," in Eli Ginzberg, ed., Technology and Social Change (Columbia University Press, New York, 1964).

generalists and integraters in analyses of general war because some aspects of the principal problems—allocation of resources among expensive competing weapons systems, cost of different styles of research and development, bargaining with a single opponent in a two-sided matrix, and so on—were well-suited to treatment by traditional economic concepts. In contrast, the dominant political character of limited wars, with numerous subtle questions of morale, training, motivation, cultural attitudes, and many-sided conflict, calls for a stronger representation of such skills as political science, cultural anthropology, and social psychology.

The infusion of social scientists into the personnel ranks of the advisory corporations does not mean that mathematicians and physical scientists have been supplanted. Rather, it serves to underscore the basically interdisciplinary character of systems analysis. A wide variety of professional skills is indispensable for effective systems-analysis work. It is the interdisciplinary character of the work which above all distinguishes systems analysis from operations research, and which also serves as a convenient benchmark to distinguish systems analysis from the applied social science research done at a number of university research centers. The university research centers devoted to the study of defense affairs and international relations typically are interdisciplinary only within the more narrow framework of a few social science disciplines.

Interestingly, the development toward the consideration of broad policy issues—and the changes in personnel background and training that made this possible—seems to have occurred principally in the United States. British and Canadian operations researchers continue to focus largely on the range of problems characteristic of World War II operations research. The researchers' training remains oriented predominantly toward mathematics and the physical sciences with almost no infusion of economists or other social scientists. Furthermore, the problems are compartmentalized along service lines with little or no consideration of the broad issues that cut across particular service jurisdictions. The fact that British and Canadian operations researchers have continued to function as civil servants, working within the government hierarchy, may be a significant reason why there has been no evolution toward the broader systems approach which examines fundamental assumptions of strategy and over-all defense policy.

Within the United States, the advisory corporations have not followed a standard pattern. Some have continued more or less in the operations-research tradition or else serve other very specialized advisory functions that do not involve consideration of broad policy issues. Other advisory organizations, by design or accident, have emerged with more general missions that include the analysis of broad policy problems as a principal task. The RAND Corporation has pioneered in this development and has set the pattern that other advisory organizations have increasingly sought to follow. Conceptual studies of broad issues have been an important part of the RAND research program since at least the early 1950's.

THE PLAN OF THE STUDY

The rise of a family of nonprofit advisory groups has awakened a lively controversy among certain strategic publics and has also attracted a fairly general popular interest. Some observers view the nonprofit advisory corporation as a kind of magical institution to feed advice to our policy makers. Suggestions have been voiced for the creation of State Department "RAND's" and for RAND-like structures to cope with some of the alleged defects of the National Security

Council's operations.[16] In the Brookings Institution Dedication Lectures, "Research for Public Policy," the name of RAND appeared frequently and suggestions were voiced concerning the application of RAND-like organizations to additional areas of public policy.[17] The creation of more RAND's has also been thought of as an answer to the real or fancied ills of bureaucracy.

At the same time, however, one sees a growing opposition to the advisory corporation. Fears are expressed that the government may be contracting out basic policy making functions to a nongovernmental body.[18] Others identify the defense advisory corporations with a particularly belligerent stance toward cold war issues, and see RAND as providing the rationale for the continuation of the cold war.[19] A more extreme interpretation posits the existence of a conspiracy, an interlocking "power elite" composed of the advisory groups together with industrial elements and military influences which threatens to preempt the governing authority of the country.[20] It should be recalled that President Eisen-

[16] Roger Hilsman, "Planning for National Security: A Proposal," *Bulletin of the Atomic Scientists* 16:93-96, 112 (March 1960), and *Organizing for National Security*, Hearings before the Subcommittee on National Policy Machinery, U.S. Senate, Committee on Government Operations, 86th Cong., 2nd Sess., 1960, parts 1-7, esp. the testimony of James Perkins in part 2, *Organizing for National Security: Science, Technology and the Policy Process*, pp. 239-268.

[17] *Research for Public Policy*, Brookings Institution Dedication Lectures, (The Brookings Institution, Washington, D.C., March 1961), pp. 25, 63, 64, 65, 93.

[18] Witness, for example, the exchange between Senator Henry Jackson and Dr. Herbert York, then Deputy Director of Defense Research and Engineering, in *Organizing for National Security*, part 2, April 1960, p. 401.

[19] Saul Friedman, "The RAND Corporation and Our Policy-Makers," *Atlantic Monthly* 212:61-68 (September 1963).

[20] C. Wright Mills, *The Power Elite* (Oxford University Press, New York, 1959) and *The Causes of World War III* (Simon and Schuster, New York, 1958); Fred J. Cook, *The Warfare State* (Macmillan, New York, 1962); and Irving L. Horowitz, *The War Game: Studies of the New Civilian Militarists* (Ballantine Books, New York, 1962).

hower also, in his January 1961 farewell address to the nation, warned against the danger of the "acquisition of unwarranted influence . . . by the military-industrial complex."[21] He feared that an undue dependence on science and scientific research as an aid to policy formation could lead to a situation in which "public policy could itself become the captive of a scientific-technological elite."

Still others, especially in Congress and in the career service, view the nonprofit corporation as simply a means to evade civil service salary limitations. On 23 June 1961, the Committee on Appropriations in House Report No. 574 noted that "to a considerable extent the use of contracts with non-profit organizations is merely a subterfuge to avoid the restrictions of civil service salary scales."[22] The committee warned that the government was "moving toward a chaotic condition in its personnel management because of this practice." The committee went on to declare that "some hard decisions must be made in regard to this mushrooming phenomenon before tremendous injury results to vital Defense programs and programs of other Departments and agencies of the federal government."[23] The bases for career service fears of the nonprofit corporation are clear enough, but congressional attitudes are more subtle and require more explanation. Some have suggested that congressmen often are favorably disposed to government contracting because this enables them to build up a constituency and a web of interests in the home district in some sense dependent upon

[21] Quoted in *The New York Times*, January 22, 1961, p. 4E. See also the authorized interpretation of this statement by the President's Special Assistant for Science and Technology, Dr. George B. Kistiakowsky, in *Science* 133:355 (1961).

[22] Report No. 574, House of Representatives, 87th Cong., 1st Sess., Committee on Appropriations, to accompany H. R. 7851, June 23, 1961, p. 53.

[23] *Ibid.*

their good offices with the federal contracting authorities.[24] Probably a major drawback with many congressmen, however, is the fear that contract employees are much less amenable to effective congressional oversight than civil service personnel.

Another source of discontent centers in various industrial circles. A comment by Ralph J. Cordiner is typical: "However generous their motives, these nonprofit organizations are usurping a field traditionally served by private consulting firms and producer companies, and hence are little more than a blind for nationalized industry competing directly with private enterprise—on a subsidized, non-taxing basis."[25] Although the brunt of the industrial reaction has been directed at the systems-engineering and technical-management corporations like Aerospace and MITRE, there has typically been little discrimination among different kinds of nonprofit organizations. The advisory corporations have thus been caught up to some extent in the industrial opposition to the "nonprofits."

Interest in the area of government contracting for research and development (R & D) had reached such a point by the end of 1961 that President Kennedy created a high-level study group, headed by David E. Bell, then Director of the

[24] Vernon Van Dyke, *Pride and Power: The Rationale of the Space Program* (University of Illinois Press, Urbana, 1964), chs. xii and xiii.

[25] "Competitive Private Enterprise in Space," in *Congressional Record*, June 2, 1960, vol. 106, p. A4719. See also the "Industry Bell Report" prepared by a committee under the chairmanship of Helge Holst, Treasurer and Corporate Counsel of Arthur D. Little, Inc., in *Systems Development and Management*, Hearings before a subcommittee of the Committee on Government Operations, House of Representatives, 87th Cong., 2nd Sess., part 1, pp. 339-402, appendix II, *Report to the Director of the Budget on Operation and Management of Research and Development Facilities and Programs, Analytical and Advisory Services, and Technical Supervision of Weapons Systems and Other Programs for the Government In-House and by Contract*, April 17, 1962, and Holst's testimony in the Hearings (hereafter referred to as the Holifield Subcommittee Hearings), part 1, pp. 79-101.

Bureau of the Budget, to review the subject. The Bell Report, released to the public on 30 April 1962, devoted considerable attention to the nonprofit advisory corporation.[26] This type of institution was viewed as a new phenomenon brought into being by the government's need for advisory services as an aid to policy formation in a complex technical era.

My general purpose in this book is to help clarify this new and inadequately explored area of what Don K. Price has termed "federalism by contract."[27] The special concern is to analyze the role of the nonprofit advisory corporation, to help clarify the rationale for its use, and to throw light on the broad problem of using our intellectual resources effectively in defense-policy formation. The approach will differ considerably from that adopted by the Bell Report. Instead of a broad look at the whole spectrum of government contracting for scientific research, technical services, and development and procurement activities, I will look closely at one part of that broad picture. The advisory corporations that normally remain "paper pure" will be my special focus. What "paper pure" means was well articulated by RAND President Franklin R. Collbohm in testimony before a congressional committee when he said that RAND "is engaged primarily in long-range research and analyses . . . as an aid to strategic and technical planning and operations. We have no laboratories in the usual sense. We do not manufacture hardware or components for sale. Normally we do not engage in design work, and are not directly concerned with the evaluation of specific products. We do not act as systems engineers, as that term is usually used in industry. Finally,

26 *Report to the President on Government Contracting for Research and Development* [The Bell Report], reprinted in the Holifield Subcommittee Hearings, part 1, pp. 191-337.
27 Don K. Price, *Government and Science* (New York University Press, New York, 1954), ch. iii, "Federalism by Contract."

RAND does not engage in the technical direction of any programs other than its own."[28] I will approach the task of analysis by undertaking a case study of perhaps the best known of the nonprofit advisory corporations: The RAND Corporation. Where useful, I make comparisons between RAND's experience and that of other defense-advisory organizations.

The reader should bear in mind that there are some salient differences in internal structure, in staffing, and in function among the various defense-advisory organizations. Although RAND is often considered an archetype by which a class of organizations is defined, RAND's uniqueness in some respects deserves emphasis. Among the advisory organizations, for example, RAND alone has seen a number of its "alumni" rise to influential positions within the government. RAND is also probably unique among the advisory organizations in the range, depth, and general quality of its professional staff. Nonetheless, there are sufficient similarities in purpose, mission, mode of operation, and relationship to client to permit one to combine the various advisory organizations into a common category for purposes of analysis. The case study method used here thus allows for depth of treatment, and at the same time affords some generalization about a "class" of institutions. An effort is made throughout to distinguish clearly for the reader when unique aspects of RAND's experience are discussed, and when RAND's experience is discussed as illustrative of the advisory organization's common problems and opportunities.

The term "advisory organization" as used in this book will refer generally to those institutions working on a more or less "permanent" basis for a defense agency or agencies. The research institution that performs a variety of *ad hoc* assign-

[28] Holifield Subcommittee Hearings, part 3, p. 920.

ments for diverse clients, government and/or industrial, will be for the most part outside the scope of this study. Similarly, the individual consultants and the special study committees used extensively by many defense agencies are generally excluded from consideration here. I will not be concerned with the nonprofit corporations, such as Aerospace and MITRE, that provide managerial assistance to Air Force commands in the technical direction of other contractors' work. I will not be concerned with the System Development Corporation except insofar as it plays a part in RAND's history. Research centers at academic institutions that are engaged in developmental, experimental, testing, or other "hardware" work for defense agencies also will fall outside the scope of this inquiry, although some attention is devoted to the university research centers engaged in applied social science research. Finally, the advisory unit located within a large industrial enterprise, or other profit-making advisory institution, is touched on only tangentially.

GENERAL ASSUMPTIONS

Because the debate on the role of the advisory corporations has so often been marked by acrimony and confusion, it is important to state the general assumptions on which this book rests. A principal assumption is that one should view the emergence of this novel type of institution in terms of science's growing impact on both the structure of government and the content of public policy. Science and technology have had one of their most dramatic impacts on public policy in the area of national defense. The military establishment, traditionally conservative in moving toward the adoption of new weapons and methods, has been forced to convert itself into a center for continuous technological innovation. This has meant a marked increase in civilian influence in areas

20

hitherto regarded as the professional military's exclusive province. For the technical skills needed to foster a high rate of innovation have not been obtainable, or obtainable in sufficient numbers, within the military itself.

Furthermore, the military has found that it could not simply lay down requirements for instrumentalities of war based on known military needs. The advent of a research-based technology and the linkage of technology with basic science has meant that change could now represent qualitative changes in the "state-of-the-art" instead of merely incremental improvements on already existing technologies.[29] The military thus could not obtain the assistance of civilian scientists and technicans merely on a junior partnership basis with civilian brainpower directed to "produce" new ideas for an advanced system as the Army arsenals in past times had produced a new rifle in response to a specified military requirement. The partnership between civilian scientists and the military in the national defense effort necessarily had to be a full one. Civilian expertise penetrated even into the traditionally sacrosanct inner councils of military planning.

The military services, in brief, recognized that they needed the assistance of civilian scientists, technologists, and analysts to operate a modern defense establishment—and they were forced to obtain this assistance to a certain extent on the scientists' terms. One consequence was that the military services began to adapt their internal procedures to provide an environment conducive to creative innovative activity. The

[29] Warner R. Schilling's comment is representative: "Until about the turn of the century, military technology, like industrial technology, generally developed independently of advances in basic scientific knowledge . . . What has transformed the relationship between science and war has been the fact that in the twentieth century the development of technology has become increasingly dependent upon advances in basic knowledge about the physical world." "Scientists, Foreign Policy, and Politics," *American Political Science Review* 56:288 (June 1962).

military "technologist" emerged as a distinct type with important responsibilities and wide influence. With the introduction of nuclear weapons and missiles, ". . . the military seems to have been almost converted into a giant engineering establishment."[30] A related consequence was that the military often found it expedient simply to purchase the services of civilian institutions engaged in scientific and technological work through the device of the administrative contract. This meant that large numbers of civilians in private institutions would have access to sensitive information and would share in the functions performed by the military. The net result is that there has been a certain "civilianizing" of the military in the scientific era. This process, I believe, is irreversible. There can be no return to a simpler age when the professional military could manage alone the business of national defense.

Conversely, the military's influence has penetrated into many areas hitherto regarded as civilian in nature. The military has a strong interest in the strengthening of science and the nation's educational standards in general. Defense expenditures have supported basic science and numerous more applied research activities at educational institutions. Inevitably, this has influenced the character and administration of the modern American university.[31] The nation's military strength also depends upon (and in turn defense expenditures help to foster) a flourishing economy and a thriving industrial community. It is a commonplace that the Department of Defense and its subordinate branches have become the employers of large numbers of civilian workers—a total, in fact,

[30] Morris Janowitz, *The Professional Soldier* (Free Press, Glencoe, 1960), p. 21.

[31] Charles V. Kidd, *American Universities and Federal Research* (Harvard University Press, Cambridge, Mass., 1959); James McCormack and Vincent Fulmer, "Federal Sponsorship of University Research," in American Assembly, *The Federal Government and Higher Education* (Prentice-Hall, Englewood Cliffs, N.J., 1960), ch. iii.

greater than the entire federal government of a generation ago.[32] Military agencies purchase a tenth of the entire U.S. gross national product and some expect this proportion to increase to a seventh during the next twenty years.[33]

The old concept of the battlefield as a sharply defined area where military forces oppose each other has given way to the modern notion of a whole nation involved in the military effort. With the advent of modern weapons and delivery systems, the civilian population shares the risks of strategic destruction along with the military. It has become apparent also that "military" and "foreign" policy are interdependent.[34] In short, just as the nuclear age has witnessed a certain "civilianizing" of the military, there has been a corresponding expansion of the military's influence which has blurred the traditional sharp distinction between civilian and military.

It seems clear (much to the distress of old-line military traditionalists and anti-military liberals alike) that this trend is not likely to be reversed in the foreseeable future. Even if a considerable measure of "disarmament" were achieved between the major powers, it is not likely that the traditional pattern of civil-military relations could be restored. The presence of modern weapons in national arsenals has introduced lasting changes in international relations and in the military's position within society. As Robert L. Heilbroner has summarized:

[32] Robert L. Heilbroner, *The Future as History* (Grove Press, New York, 1960), p. 66.

[33] Committee for Economic Development, *Problems of United States Economic Development* (New York, 1958), I, 29, essay by Dr. Simon Kuznets.

[34] Henry A. Kissinger, *Nuclear Weapons and Foreign Policy* (Harper, New York, 1957); W. W. Kaufmann, "Force and Foreign Policy," in W. W. Kaufmann, ed., *Military Policy and National Security* (Princeton University Press, Princeton, 1956), pp. 233-267; Dean Acheson, *Power and Diplomacy* (Harvard University Press, Cambridge, Mass., 1961); and Samuel P. Huntington, *The Common Defense: Strategic Programs in National Politics* (Columbia University Press, New York, 1961).

But even a considerable relaxation of tension will hardly permit us to return to the pre-atomic subordination of military affairs. As we shall have cause to see, the world will be an exceedingly dangerous place to live in over the foreseeable future. The outbreak of small wars is virtually a certainty, and some of these will have the potential to develop into large wars. The likely extension of nuclear armaments to secondary powers in the near future, and possibly to tertiary powers in the not too distant future, will all add uncertainty and risk to the prospects for international history. Thus it is extremely unlikely under the best of circumstances that we shall permit our highly trained cadres of professionals in nuclear or conventional war to dissipate their skills, their organization, their effectiveness. Nor is it likely that we would allow the facilities for the production of war goods to remain unused, or even unbuilt. The position of high importance, prestige, and sheer magnitude or organization which the military now occupy in *every* atomically armed industrial power must be regarded as a historic result of the impact of the new technology of war upon modern society.[35]

This study assumes, therefore, that the United States will have a substantial military establishment for some time to come even though an atmosphere of *detente* may considerably reduce international tensions. As a corollary to this, it is also assumed that the United States will seek to maintain a technological base—or capacity for producing modern instruments of warfare—even though there may be fewer new weapons systems actually introduced into the nation's inventories in the coming decade.

Given these conditions, it will remain important for policy makers at various levels to have research and analytic work performed on major issues. High-level civilian officials will need a source of advice apart from the professional military so that a basis for critical evaluation of military proposals can be maintained. Indeed, the concept of a "countervailing expertise" to the professional military's judgment will in-

[35] Robert L. Heilbroner, *The Future as History*, p. 68.

creasingly be central to the task of effective civilian control of the military. Much of the traditional thinking on the subject of civilian control "has been rooted in the specter of the 'man on horseback' rather than on the relevant framework of the twentieth-century problem of the relation of the expert to the politician."[36] It is a curious irony that the military services themselves, in sponsoring organizations like RAND, have greatly strengthened the civilian's role in defense management and policy formation.

From the professional military's point of view, a community of civilian analysts will remain important as a means for helping acquaint the civilian policy makers with the military facts of life. Assistance in assessing what sort of technological base to maintain, and how an evolving technology may affect military doctrine, operations, and organizational requirements, will also continue to be useful to the professional military.

Given the continued presence of firms selling military instrumentalities to the government (and/or "peace" technologies), all defense decision makers will benefit from a source of advice divorced from the market on the feasibility and cost of advanced systems. And decision makers at all levels will benefit from systematic efforts to reduce the uncertainty in difficult choices, to define how national objectives can be translated into hardware, to assess the consequences of pursuing different objectives, and to provide some basis for identifying relevant national objectives in a future military environment where complexities are so great that the decision maker's unaided intuition can hardly comprehend the dimensions of his task.

It is in these terms that the role of RAND and similar

[36] Harry Coles in Coles, ed., *Total War and Cold War: Problems in Civilian Control of the Military* (Ohio State University Press, Columbus, 1962), p. 3.

advisory institutions must be understood and evaluated. How effectively does RAND (and the related institution) perform an advisory function for defense clients? What are the special advantages and disadvantages of the RAND-type structure? Would an alternative institutional arrangement be preferable? How did RAND come to be a "nonprofit corporation?" What does "nonprofit" mean in this context? Does nonprofit status effectively insulate RAND from market pressures and conflict-of-interest difficulties? Are there special problems in attracting and keeping a staff of a high order? How does sponsorship by a military service affect the advisory operations? Can the advisory institution successfully serve both a military serve and agencies at the Office of the Secretary of Defense level? What would be a "model" of an effective advisor-client relationship with reference to formulation of work projects, communication of analytic findings, publication of unclassified reports, and the general "independence" that the advisory institution should enjoy? These are some of the relevant questions to ask. And facing up to such questions is no idle exercise. A task of growing urgency for policy maker and scholar alike will be to understand the processes by which technical expertise can aid political judgment on pressing public issues where science and technology are deeply involved.

Several final clarifications are necessary. It is not my intention to be concerned primarily with substantive issues of national security policy. The influence of RAND on specific policy decisions has often been greatly exaggerated by friend and critic alike. It is fanciful to suppose that the activities of RAND and organizations like it provide the main impetus for a continuation of the cold war. The clash of national interests, the rise of defense-related industries on a large scale, tension and instability in international politics, and

the growth of substantial military establishments in all industrial nations can hardly be ascribed solely to alleged
machinations and manipulation of thought processes by
RAND and other advisory bodies. An opposite statement is
more accurate. These "objective" factors, combined with the
injection of an accelerating technology into human affairs,
formed an environment in which analytic advice on complex
technical-military-political issues was very useful. Conspiracy
theories seldom capture the richness and complexity of social
processes. The intricate interplay between advisor and client
in a dynamic setting is greatly oversimplified by theories
which impute an extraordinarily powerful and sinister behind-
the-scenes role to the RAND-type institutions.

Similarly, it strains the imagination to give the advisory
community major credit for "successful" defense policy. The
advisor's influence is necessarily limited to a portion of the
full spectrum of problems and choices that the policy maker
faces. Intellectual resources are in finite supply and the research organization must be selective in choosing topics for
study. And normally analytic skills can be profitably applied
only to those issues where time permits debate, discussion,
gathering of data, and the iterative processes of scholarship.
More important, there is a crucial conceptual point that
should be stressed. As noted earlier, there is always some gap
between the analytic product and the "real world" constraints within which the policy maker must operate. In
principle, no amount of research or advice can ever definitively treat all of the variables that go into a complex decision.
Research and advice can supplement, but can never replace,
the exercise of executive judgment. The typical analytic
product thus seldom emerges with simple guides to action
on broad problems (although on occasion a study may have
some clear-cut policy implications). The conceptual study

27

seeks mainly to enhance the policy maker's understanding of a difficult issue. As such it is only as useful as the wisdom and discretion of the officials who must understand, evaluate, and apply it to a context which normally includes numerous "extra-analytic" considerations. Although the advisory process has been institutionalized in defense-policy formation, this has not meant that at the focal point of decision the responsible official has abdicated his ultimate authority. Successes of U.S. defense policy should be attributed principally to the imagination, resourcefulness, and steadiness of purpose shown by our policy makers (and, one may add, to their ability to judge when, and how, and with what qualifications technical expertise should be translated into policy).

It is therefore neither possible nor desirable simply to add up in some fashion examples of RAND research and to assess them in terms of whether they lead to "good" or "bad" defense policies. What is principally at issue, rather, is RAND's effect on the "style" of defense decision making. The advisory organizations have made their presence felt in the defense establishment chiefly by influencing the way in which policy makers approach their task, define alternatives, and assess criteria for choice. One should ask whether there are defects, systematic biases, errors of concept or execution in analytic work by the advisory corporations. But it is inappropriate to evaluate the advisory corporation's work solely on the basis of whether or not one favors specific policy choices ultimately made by responsible officials. The research product can be misunderstood, poorly or uncritically evaluated, misapplied, misused, ignored, or made too much of or too little of at numerous points in the decision process. The advisory corporation has some responsibility to help prevent misapplications of its research products, and quite properly should strive to obtain a wide hearing for its views in the client agency. However, it distinctly should not be the advisory corporation's function to exercise decision-making power itself. There is probably little danger at present that the advisory body, in throwing up approaches to problems for the policy maker's consideration, will acquire power for itself. The real difficulties will be more subtle questions of framing an institutional environment that will permit coherent, unified yet adaptable national policies, avoiding conflicts-of-interest of an unusual and little-understood nature, developing better criteria for assigning research tasks to particular organizations, and evolving leadership strata (both career and political) within the government that can keep a steady and guiding hand on the advisory apparatus and maintain a clear central direction in policy. An understanding of the advisory corporation's role should help provide a better basis for the efficient yet responsible use of our intellectual resources in defense-policy formation. The experience of the defense agencies with the RAND-type advisory corporations may also serve as useful guidelines for other public policy areas. The administration of policy-oriented research activities will clearly become a subject of increasing practical and theoretical interest as policy choices become ever more complex and policy makers are less and less able to rely on conventional wisdom alone as a basis for decision.

► II ◄

THE BACKGROUND AND ORIGINS
OF PROJECT RAND

> The Contractor will perform a program of study and research on the broad subject of intercontinental warfare, other than surface, with the object of recommending to the Army Air Forces preferred techniques and instrumentalities for this purpose.
>
> —*letter contract of 2 March 1946 establishing Project RAND*

The previous chapter began to lay the necessary foundation for understanding the emergence of RAND by noting the impact of science and technology on modern warfare. The military establishment, long traditionalist in nature, was forced to convert itself into a center for continuous technological innovation. Warfare became so technically complex that the military required extensive aid from civilian scientists and technologists; in consequence, civilian influence penetrated into numerous areas once considered the exclusive province of the professional military. In this context one can see how the concept of a civilian advisory institution could arise and be carefully considered by military planners. Some kind of scientific advisory role, in short, came to be a "functional requisite" for the modern military establishment.[1]

---

[1] On the notion of a "functional requisite," see Marion J. Levy, *The Structure of Society* (Princeton University Press, Princeton, 1952).

This is not to suggest any teleological or functionalist interpretation of social phenomena that sees the emergence of the RAND-type advisory institution as "necessary" or foreordained in some sense. Recent advances in sociological theory have largely discredited the notion that any component part of the social system can be explained in terms of the system as a whole and its "functional" role in the system's operation.[2] An institution or other element within a social system does not necessarily contribute to the system's continuity or survival. It may be "dysfunctional"; it may contribute to the system's eventual downfall; or it may be functional with respect to one part of society and dysfunctional with respect to society as a whole. It is also possible that an element within a system may have no significant consequences and simply be a functionless form. Social systems generally have come to be viewed as exhibiting possible tensions and discordant elements, with the important theoretical implication that an existing institution or structure can be destructive, irrelevant, or replaceable by some other unit.

It would be a mistake, therefore, to say simply that an evolving military technology "required" the creation of a RAND. What can be said is that the complexity of modern warfare made it highly desirable, and probably essential, for the military to obtain scientific advice on numerous aspects of planning, operations, and weapons-development policies. But how was this advice to be obtained? One can imagine a variety of responses that could serve to integrate the fruits of scientific research closely with military planning. Civilian scientists could be absorbed directly into the military organization; special government research organizations could be established; the military could drastically alter personnel practices

2 Wilbert E. Moore, *Social Change* (Prentice-Hall, Englewood Cliffs, N.J., 1963), esp. pp. 6-21.

to provide large numbers of scientifically trained officers; industrial firms could supply scientific advisors; or university research centers could provide advisory services under a contractual arrangement. One can also imagine a number of non-adaptive responses inasmuch as social structures are by no means self-adjusting. Even when the need for innovation is widely perceived, an organization may either fail to innovate or else may innovate unsuccessfully.

Some responses, however, can probably be considered *prima facie* less likely than others. Civilian scientists, for example, probably could not have been brought directly into the military establishment in sufficient numbers after the war because many scientists had chafed under wartime restrictions and were deeply suspicious of military supervision. The widespread recognition of the need for scientific assistance among progressive military leaders made it unlikely that the military would simply revert to the prewar neglect of science and technology in military planning.[3] In retrospect, it does not seem unusual that the military services, following the wartime precedent set by the Office of Scientific Research and Development (OSRD), would contract with outside institutions and individuals to obtain technical advisory services.

These comments do not, however, go very far toward explaining the peculiar prominence of the nonprofit advisory corporation. Why did RAND and other advisory organizations come to assume this particular form? Why did RAND

[3] As General Eisenhower, for example, then Chief of Staff, told the Army in 1946: "The future security of the nation demands that all those civilian resources which by conversion or redirection constitute our main support in time of emergency be associated with the activities of the Army in time of peace," and he went on to advise the Army to contract extensively for scientific and industrial services (*Memorandum for . . . General and Special Staff Divisions*, "Scientific and Technological Resources as Military Assets," 30 April 1946), quoted in Don K. Price, "The Scientific Establishment," *Proceedings of the American Philosophical Society* 106:235 (June 1962).

evolve one pattern of relationships with the client agency and not some other? The answers to these questions provide the materials for an intriguing piece of administrative history. More important, perhaps, the story of RAND's development contains some lessons for the broad question of how our society can organize to use analytical skills in policy formation. Many of the questions now being raised about the proper institutional setting to carry out policy-research tasks have arisen in RAND's history. Some RAND practices, while not necessarily authoritative precedents, offer useful lessons on problems commonly faced in the advisor-client relationship. In other respects, RAND's experience has been so unique and idiosyncratic that it probably offers only negative lessons. For example, one of the most arresting features of RAND's development—the unusually high degree of independence and freedom from sponsor control or domination—is probably in no small part attributable to serendipity. Chance, circumstance, historical accident have all played an important part in RAND's history. Thus it might not always be possible to duplicate by design the conjuncture of accidental circumstances that contributed to the favorable environment for RAND's development. But a look at RAND's history can at least suggest some of the right questions and possible answers for those who must make decisions on how to organize, staff, administer, and make effective use of the policy-research organization.

WORLD WAR II AND OPERATIONS RESEARCH

For the first phases of the RAND story, it is necessary to look back to World War II and to the variety of operations-research activities carried out then. As previously noted, the initial applications of operations-research techniques came in the fields of radar and anti-submarine warfare in the early phases of the war. The creation of the Antisubmarine War-fare Operations Group (ASWORG) in March 1942 marked the first effective organization devoted to operations research to be established in the United States.[4] Not long after, the Army Air Forces became interested in operations research (or, as the British called it, operations analysis). Operations-analysis sections were established, composed principally of civilians with physical science and mathematics backgrounds, to be attached to Army Air Force commands overseas.[5]

Also, by this time various *ad hoc* groups of specialists from the OSRD complex and from the office of Dr. Edward L. Bowles, consultant to the Secretary of War, were beginning to go into the "field" on special missions to help the services use their new weapons and equipment efficiently. An im-portant part of this work could be considered operations re-search, though on occasion these missions served mainly a training function. Operations-research techniques at this time, it is important to note, were designed to do primarily one of two things: (1) They either analyzed the problems

[4] James Phinney Baxter, *Scientists Against Time* (Little, Brown, Boston, 1947), p. 405, and Samuel Eliot Morison, *The Two-Ocean War* (Little, Brown, Boston, 1963), pp. 122-129. Even earlier, however, in the fall of 1941, a small informal group was started by Dr. Ellis Johnson at the Naval Ordnance Laboratory, with encouragement from visiting British scientists, to study mine warfare and counter-measure operations.

[5] However, the first civilian Air Force operations analysis section was headed by a lawyer—John Marshall Harlan—who later became an Associate Justice of the Supreme Court. One explanation of this apparent anomaly is that when U.S. officials queried Professor P. M. S. Blackett about British experience he said that the British had tried practically every profession *except* corpo-ration lawyers. His hearers misunderstood this remark as a recommendation and set out to enlist a prominent lawyer to head the first section (from an unpublished paper by Dr. Lynn H. Rumbaugh, Research Analysis Corporation). This story, however, is probably apocryphal. More likely, lawyers were probably thought to have analytical minds well suited to supervising operations research activities. The first such group (the Operations Analysis Division) formed at the Washington level was also headed by a lawyer, Colonel (later Brigadier General) W. Barton Leach.

involved in making new weapons and techniques useful to the military. The advent of radar and its integration into an operational military context is the classic example of this category of operations research. Or (2) they dealt with tactical military "operations" and consequences of the interplay of new weapons systems and advanced technologies in a tactical military setting. Anti-submarine warfare is the outstanding illustration of this kind of early operations research.[6]

Later in the war, however, operations research achieved a broader use as well in direct participation in military planning. For the first time in history, civilians were brought into the inner councils of military planning. Warren Weaver expressed very well the import of this development in his "Opening Plenary" remarks to RAND's Conference of Social Scientists in September 1947:

> They [the military] were quite willing to accept civilians on a certain service level in the past. They used to say "We like to have you around, and if you are awfully smart we will ask you questions and you will answer them as well as you can; but then we will go into another room and shut the door and make our decision." That, in the past they were quite willing to do. Now, however, they want us in the backroom with them. They want to talk over the really fundamental questions, and they are actually admitting civilians at the planning level. That, I think, is very significant.[7]

One of the first examples of this direct civilian participation in military planning occurred in the B-29 Special Bombardment Project. It is worth pausing briefly to examine the origins and results of this project. For it was out of this experience that the Project RAND idea began to emerge; and

[6] For a description of the early operations-research activities, see Philip M. Morse and George E. Kimball, *Methods of Operations Research* (John Wiley, New York, 1951).

[7] *Conference of Social Scientists*, Report 106, The RAND Corporation, (September 1947), p. 5.

here the personal contacts were formed which eventually led to the Project RAND contract. The idea for the Special Bombardment Project originated in early 1944 with Dr. Edward L. Bowles, special consultant to General H. H. "Hap" Arnold, Army Air Force Chief. Bowles, one of the cluster of leading scientists and engineers brought into the government through the OSRD mechanisms to aid in the war effort, transferred early in 1942 from the OSRD to the War Department where he became a special consultant on science and technology to Secretary of War Henry L. Stimson and General Marshall. There he built up his own staff, a kind of OSRD in miniature, within the War Department. By the late stages of the war, he had developed a highly competent team that was working closely with the OSRD and making a significant contribution to the war effort.[8]

From 1943 on Bowles worked increasingly with General Arnold assisting the Army Air Forces in adapting the nation's air might to the stream of new devices and techniques coming from the OSRD scientists and technicians. Bowles gained the confidence and respect of Arnold, and enjoyed unusual leverage as a gadfly advisor to Air Force officers. Early in 1944 he decided to strike out on a somewhat novel undertaking. Convinced of the need for even closer cooperation

---

[8] For a brief description of Bowles' operations, see Baxter, *Scientists Against Time*, pp. 33, 408; Lincoln R. Thiesmeyer and John E. Burchard, *Combat Scientists* (Little, Brown, Boston, 1947), pp. 257, 262 and 262n; Elting E. Morison, *Turmoil and Tradition: A Study of the Life and Times of Henry L. Stimson* (Houghton, Mifflin, Boston, 1960), p. 563; and Henry L. Stimson and McGeorge Bundy, *On Active Service in Peace and War* (Harper, New York, 1948), pp. 468, 510. Among Bowles' main achievements were the establishment of the Sea-Search Attack and Development Unit (SADU) at Langley Field in 1943, which played an important role in combating the submarine menace on the Atlantic seaboard; the assignment of radar-equipped search aircraft to the Far Eastern theater; and the rapid transfer in 1944 of "gunlaying" 584 radar units which helped Great Britain combat the "buzz bomb" menace.

between the scientists and the military, Bowles decided to attempt to integrate the two more closely by bringing scientists directly into the Army Air Force planning process at the Washington staff level. (Previously, much of the technical advisory work had been done in the field and within the context of a specific operational or tactical situation.) The strategy was to organize a study team to undertake an analysis of ways to improve the effectiveness of the B-29 as a strategic bomber. The study team was to work closely with Army Air Force officers, receive thorough briefings on Army Air Force plans, and have complete access to all other necessary information. Accordingly, Bowles summoned the chief engineer of the Boeing Company, Edward Wells, to the War Department as a consultant to execute the study. Shortly thereafter, Bowles brought in the chief engineer of the Douglas Aircraft Company, Arthur E. Raymond, and his aide, Franklin R. Collbohm, to assist in the project.

The project's success exceeded even the most optimistic expectations. The report of the Wells-Raymond-Collbohm team showed that the range, speed, and bomb load of the B-29 could be greatly enhanced by stripping away much of the heavy defensive armor that gave unnecessary weight to the airplane.[9] The heavy defensive armor, the report demonstrated, served little useful purpose because without the armor the B-29 had a speed superior to that of any known Japanese fighter. The report recommended, however, that a tail gun along with appropriate rear armor be retained to guard against possible one-pass attacks from enemy fighters diving from above. The Army Air Force was receptive to the

[9] Special Bombardment Project Interim Report, October 7, 1944. Declassified. Obtained from the personal files of Dr. Edward L. Bowles. The personal papers and letters of Dr. Bowles are the best source of documentation on Project RAND's founding. I wish to record my special gratitude to Dr. Bowles for granting me access to his personal files.

study team's suggestions. Subsequently, a whole wing of B-29's was equipped according to the specifications laid out in the study and was used to great effect in the Far Eastern theater in the closing phases of the war.[10]

BIRTH OF PROJECT RAND

Meanwhile, starting in late 1944 a number of informal conversations were held in War Department and OSRD circles about the problem of maintaining an effective partnership between scientists and the military once the war was over. As victory drew near, numerous scientists were growing restive under the discipline of the war effort. After the close of hostilities, it was apparent that many scientists would eagerly return to industry and to the academic community. Many people feared that the effective wartime partnership between scientists and the military could not be maintained in peacetime. General Arnold, who had seen the Germans develop jet aircraft in the final stages of the European war that might have stopped the whole Allied air offensive had they concentrated on fighter rather than bomber production, was among those deeply concerned with the problem at an early stage.[11]

[10] The "pay off" was recorded in a August 8, 1945 cablegram from General Spaatz to Generals Norstad and Twining: "Have just been shown photographs of night attack on 5/6 August against the Ube Coal Liquification Co. by 315th Bombardment Wing. This operation, using blind bombing techniques, shows precision bombing which has not been exceeded by visual methods in daytime. Please convey my commendations to all concerned." (Source: Bowles papers.)

It should be noted that, consistent with the view of research as an important yet subordinate aid to decision making, the B-29's were actually used for a somewhat different purpose than the Special Bombardment study had suggested. Low-altitude night bombing, instead of high-altitude daylight bombing, was the use to which the stripped-down B-29's were put.

[11] See Gen. Arnold's testimony in *Department of Armed Forces and Department of Military Security*, Hearings before the Committee on Military Affairs, U.S. Senate, 79th Cong., 1st Sess., on S. 84 and S. 1482, October, November, December 1945, pp. 67-96.

Dr. Bowles was another who had given much thought to the need for some kind of postwar mechanism to retain the services of scientists of all sorts in the national defense effort. Numerous others in military, industrial, and scientific circles were also vitally interested, including Paul Scherer in an important position as head of the OSRD Transition Division. As the war drew to an end the problem received increasing attention.

One suggestion that floated around Washington at this time was the idea that the government should create a general "procurement agency" for buying brains instead of physical items. Other ideas stressed the feasibility of contract relationships with private institutions, which had been a trademark of OSRD operations. In this general atmosphere, the idea emerged of a contract with a private organization to assist in military planning, and particularly in coordinating planning with research and development decisions. Thus the concept of Project RAND began to emerge in nascent form in midsummer 1945.

It is difficult to pinpoint a single place where or a single individual with whom the idea originated; it gradually took shape in informal discussions among men in the War Department, the OSRD, and industry. Two groups, however, must be singled out for particular mention. One was the group that centered around Bowles in the War Department, including L. J. Henderson, Jr., Julius Stratton, and David Griggs; and the other was a group at the Douglas Aircraft Company led by Bowles' former consultants, Raymond and Collbohm, who had in the meantime resumed their normal duties with the Douglas Aircraft Company. The initiative for the idea came in considerable measure from the Douglas people, and Bowles took the main responsibility for nurturing the idea and steering it through difficult obstacles on the Washington front. By the end of the summer of 1945, Collbohm had made several trips to Washington to discuss the idea in tentative form with Bowles and others, including General Arnold and Secretary of War Patterson. Finally, in the fall of 1945, Collbohm came to Washington with a proposal for the Douglas Company to house a civilian group that would assist the Army Air Force in planning for future weapons development.

It should be noted here that there is no evidence to support the view, which subsequently gained some currency, that the company reluctantly accepted the Project RAND contract at the government's urging out of lofty patriotism. It seems clear that the company in a general sense sought the contract, though the specific timing and initiative for activating the project in the fall of 1945 rested principally with General Arnold and Dr. Bowles. Later, when Project RAND threatened to become a business disadvantage, the Douglas Company's initial enthusiasm cooled and it sought to back off from sponsorship of the project. The award of a large air transport contract in 1947 to the Boeing Company, which the Douglas people had expected would come to them, helped to crystallize the feeling within Douglas that Project RAND was a liability because the Air Force was extremely anxious to avoid the appearance of giving preferential treatment to Douglas.

General Arnold was impressed with the Collbohm proposal, and decided it was possible to act on the matter within the Air Force itself without waiting to coordinate with the War Department.[12] He accordingly asked Collbohm to ask Douglas to meet him for lunch the following day at Hamilton Field out-

---

[12] This reconstruction of the events surrounding Project RAND's inception is based on the written documents in Bowles' files and on personal interviews with the following RAND "founding fathers" who are still alive—Bowles, Collbohm, Raymond, Henderson, Goldstein.

side of San Francisco where a final decision or at least some sort of understanding would be reached. On October 1, 1945, the group of "founding fathers" met at Hamilton Field and framed what was to become Project RAND.[13] Present at the meeting were General Arnold, Dr. Bowles, and various representatives from the Douglas Company, including Donald Douglas, Sr., Arthur Raymond, Franklin Collbohm, and Frederick W. Conant.

What happened at the Hamilton Field meeting is essentially the following. General Arnold indicated that he had certain moneys left over from his research budget for the war, and that he wanted to set aside three packets of $10,000,000 each for the study of V-1 and V-2 rocket techniques and other intercontinental air techniques of the future. Assuming a contract was to be consummated eventually, one of these packets was to be available for the Douglas Company Proposal. (The other two projects never got beyond the planning stage and were lost in the confusion of demobilization.) All present agreed on the importance of a continuing scientific effort on the techniques and devices of intercontinental air warfare. The Douglas people indicated their eagerness to move ahead with the project as rapidly as practicable since they would otherwise have a difficult time holding the requisite staff in view of the fact that many of the company's wartime projects were then being liquidated. The sense of the meeting was that the company would begin to take preliminary steps to activate the project in the expectation that a formal contract would be negotiated in the near future. The specifics of the project and the relationship between the com-

---

13 The name, Project RAND, was coined by Arthur Raymond and was intended as an acronym for "research and development." Later, as one wag pointed out, the initials were even more appropriate: they spelled out Research and *No* Development.

pany and the government were to be worked out during the contract negotiations. General Arnold delegated to Dr. Bowles the major responsibility for activating the project and working out the arrangements with the Douglas Company.

Several major considerations underlay the choice of the Douglas Company as the home for the new project. First, it was considered necessary to attach the project to a going concern (at least initially) in order to help get the project off to a proper start. The Douglas Company at the time was one of the country's biggest and most prosperous airframe manufacturers. Moreover, the Army Air Force had had successful working relationships with the company during the war, including the B-29 Special Bombardment Project which served as something of a prototype for the new project. Second, the founders thought that a university would not consider having anything to do with highly classified research projects of this kind. That this estimate turned out to be incorrect did not lessen its effect in steering RAND away from direct university ties. Third, by general agreement it was considered impossible to build such a high-talent scientific group within the government itself because of the poor salaries and inflexible personnel practices that prevailed in the civil service at the time. Furthermore, the founders felt that scientists would be difficult to recruit if the project was administered directly by the military.[14]

General Arnold also very likely felt that in Donald Douglas he had a man he knew and trusted and who could be relied on in what was after all a rather vague and uncertain venture. Arnold had known and worked with Douglas for some time, and their two families were related by marriage. One of

[14] Several of the framers were also concerned with finding a location far enough away from Washington to minimize the number of routine requests directed to the research staff. On the locational factor, see pp. 74-76.

General Arnold's three sons, William B. Arnold, had married Donald Douglas' daughter in 1943 after his graduation from West Point.

EARLY UNCERTAINTIES AND DIFFICULTIES

The Hamilton Field meeting left a number of points unsettled about the nature and purpose of the proposed Project RAND. Although all of the "founding fathers" shared the conviction that the partnership between scientists and the military must continue into peacetime, different views were held as to the exact form the partnership should assume and the precise objectives that it should pursue. In the first place, it was not clear to what extent the project would be exclusively an Air Force undertaking or whether at some future point the other services would participate. Possibly the project would be integrated within the general procurement agency for brains that was under consideration at the time. The founders generally felt that the Air Force should proceed on its own for the time being because of the enormous difficulties involved in trying to coordinate the project with the other services at this early stage. Also, they were afraid of the delay that would result if no action was taken pending formation of some statutory agency for procuring scientific skills. But the option was left open to broaden the project's sponsorship if necessary at some later date.[15]

15 Letter from Edward L. Bowles to Secretary of War Patterson, 4 October 1945. Bowles wrote in part: "There are, of course, not only contract problems to be resolved so as to make it reasonable for a commercial organization to operate on a project of this magnitude, but also the larger problem of integrating an undertaking of this broad character with other efforts in the Army, with the Navy, and other governmental agencies. The other agencies would include the Research Board for National Security, the Manhattan District Authority, and whatever body Congress may set up for general research. However, I believe it would be a mistake for us to withhold going ahead with Air Forces' plans until all those factors are harmonized. It should

A second unclear point was the exact role the Douglas Company was to play in administering the project. It was generally understood that the project was to be physically and administratively segregated from the company's regular activities. But it was not clear whether the arrangement with the company was intended to be only provisional in nature. Would other aircraft companies (and other industries) participate in the project's over-all direction? If so, how? Bowles, for example, held the notion that the project should eventually broaden into an institution supported by the entire airframe industry.[16] There was also some thought given to the idea that the Douglas Company would ultimately establish a charitable foundation supported by company funds and dedicated to the pursuit of research in aviation. Indeed, Bowles believed that Douglas had given him an informal commitment to take over the project's financing at some future date. This was one reason, among others, why Bowles grew increasingly unhappy with the way the Douglas Company was administering the project.

Finally, exactly what kind of work the project was to do was by no means clear. In fact, in one of the working notes for the Hamilton Field conference, it was proposed that the project be carried out in two stages. The first would involve only background research on instrumentalities of intercontinental warfare, and the "final phase would be negotiated after the first phase is completed and would end in the development of a physical article."[17] Even if it were presumed that Project RAND would do only background research and no development work (the pattern RAND actually followed),

be simple to integrate the Air Forces' effort later when we know better what our peacetime research organizational structure is in fact." (Bowles papers.)

[16] Bowles papers.

[17] Memorandum prepared by Arthur E. Raymond for the Hamilton Field Conference, n.d. (Bowles papers.)

the question still remained as to what kind of research would be done. Would it be research only in the "hard" sciences? Or would the social sciences be included as well? Indeed, one of the interesting questions in RAND's development is the extent to which (if at all) the social sciences were implicit in the RAND concept from the beginning. It is fair to say that there was no unambiguous agreement among the framers on the social science role. The concept of social science research may have been present in the minds of some of the persons associated with Project RAND's founding, but it was certainly not present in all the founders' minds. For the most part, the issue was probably not consciously thought about at this early stage.

After the Hamilton Field meeting, serious questions were raised within the Air Force about the adequacy of existing administrative machinery to monitor the new project (and any similar long-range research efforts that might be undertaken). Of particular concern to the Air Force proponents of the novel Project RAND concept was the danger of the procurement and materiel function dominating and stultifying the long-range research effort. General Lauris Norstad, then Assistant Chief of Air Staff, Plans, was one of the principal Air Force figures who early recognized this danger. Dr. Bowles also was particularly active in the period following the Hamilton Field meeting in stressing the need for an administrative arrangement that would allow full scope for the project's development. Norstad and Bowles were instrumental in convincing Arnold and other "liberal" Air Force officers that it would be fatal to the success of the project if it fell under the domination of Wright Field materiel people with their traditionalist and "requirements approach" type of thinking. Hence, Arnold was persuaded to establish a new office within the Air Staff above the regular staff sections to ensure high-level consideration for and enlightened administration of the new project. The man chosen for the assignment was intentionally a field commander without experience in old-line research and development work. On 1 December 1945 the new office of Deputy Chief of Air Staff for Research and Development was created, and General Curtis LeMay named its first incumbent. LeMay proved to be an important asset to RAND in its early years. He helped the project to get established on a firm footing and generally worked to safeguard RAND's independent status. It is a matter of some irony that in later years LeMay grew to be a strong critic of RAND.

Not long after LeMay's appointment, a general statement of work was produced for Project RAND. General LeMay summoned representatives from the Air Staff and the Wright Field Materiel Command to a meeting to work out a policy statement for the new research effort which, at this time, still had not yet been officially activated. There was a good deal of opposition to the Douglas proposal expressed at the meeting. As expected, particularly sharp opposition came from the delegation of materiel and procurement officers from Wright Field. They objected that what was being proposed was civilian assistance in military planning. They also expressed opposition to bringing the Air Force planning process into the research and development area. These subjects, the delegation held, were the prerogatives of Wright Field. The Wright Field delegation proposed instead a tightly controlled program telling the contractor what it should do and specifying in rigid detail a research effort leading to a missile with a range of 13,500 miles, a warhead of a thousand and some pounds, a certain number of missiles to be launched per hour, and a statement of precision of hits.[18]

18 Based on a 26 November 1946 memorandum from Edward L. Bowles to Air Force Secretary Stuart Symington and on interviews with Bowles and

Fortunately for the future of Project RAND, General Le-
May exercised the prerogatives of his office and overruled
these objections. He pointed out that the purpose of the project was not to state a requirement and tell the contractor what
to do. Rather, the contractor was to perform long-range research that might form the basis for a future military requirement. He also indicated that the new organization was to have
a high degree of freedom to carry out its research objectives.
Thus began a recurring pattern in RAND's history: RAND
was fortunate enough to have "protectors" at high levels
within the Air Force at crucial points to prevent it from fall-
ing victim to an internal power grab or from collapsing under
severe budget cuts inspired by critics.[19] One must be careful,
however, not to paint an overly melodramatic picture. Large
organizations are seldom monolithic, and it is not surprising
that RAND has always had "enemies" as well as "friends"
within the Air Force. What is noteworthy is the consistent and
surprisingly enlightened sponsorship RAND has in general
enjoyed through the years of association with the Air Force.

In any event, the statement of work which finally emerged
from the LeMay meeting was a broad and permissive program of research on "intercontinental warfare, other than
surface, with the object of advising the Army Air Forces on
devices and techniques."[20] The wording was suggested by
Bowles. Several aspects of the broad work statement are
significant and deserve emphasis. The phrase "other than

Franklin R. Collbohm. (Collbohm and David Griggs, one of the original RAND
staff members, represented Project RAND at the meeting.)

[19] Some of the "flavor" of this aspect of RAND's history was supplied me
in an interview with Lt. Gen. Donald L. Putt (USAF, ret.), formerly Air
Force Chief of Research and Development and intimately acquainted with
RAND affairs since 1949. Interview took place in San Francisco (Sunnyvale
suburb), December 20, 1961.

[20] Bowles memorandum to Symington, November 26, 1946. The statement
of work contained in the letter contract formally consummated on March 2,
1946 was almost identically worded.

surface" was included to avoid offense to the Navy. And the words "devices and techniques" strongly suggest that the research effort was supposed to emphasize the physical sciences. This lends support to the view that few people associated with Project RAND foresaw at this early stage the enormous impact that social science thinking later was to make on the character of the RAND research effort. Finally, the broad wording of the statement meant that, in effect, the new organization was given a virtual *carte blanche* to work out its own research program. On 2 March 1946, a letter contract was consummated with the Douglas Aircraft Company bringing Project RAND officially into existence. At this point, Project RAND had some four employees occupying a walled-off section on the second floor of the main Douglas building in Santa Monica.

In looking at RAND's early history, it is well to recall General Arnold's foresight in providing the initial financial support for a period of several years without insisting on any immediate and tangible results. This period of grace enabled RAND to pursue a number of blind alleys, to experiment with methods of operation, and in general to get on its feet without intense pressure for an early "pay-off." RAND at this time also enjoyed the advantage of obscurity. It had attracted almost no public attention and congressional interest in the project was nil. RAND also benefited considerably in its recruitment efforts by the fact that a number of top-flight scientific personnel who had participated in the war effort were then in the process of returning to civilian life and were available for employment in the new organization.[21] In a later day of somewhat more settled career patterns a fledgling research group would not have this advantage.

[21] There were, however, other factors which complicated recruitment, notably the Douglas connection. See notes 29 and 30 below.

Moreover, at the time, the mission of the Air Force was almost equivalent to the whole of the national security effort. The Air Force was secure in its possession of the atomic bomb and the means of its delivery, and had little reason to fear either foreign enemies or a competing sister service.[22] This meant that outstanding individuals could be more easily attracted to the project as it promised to deal with broad problems of national security. Now it would be much more difficult to attract able people to a research project geared to the problems of only one of the services. In short, RAND had a number of factors working in its favor that contributed to its eventual preeminence in the defense-analysis field. It is doubtful whether any similar research group established today could expect to begin in such auspicious circumstances.

However, it should not be thought that these advantages enabled RAND to achieve success overnight or that there was a simple and easy route followed from the start. In actuality, RAND was to encounter a number of difficulties, some minor and some of a fundamental nature, and to undergo significant growing pains in its early years. The fundamental difficulties eventually led to the separation of Project RAND from the Douglas Company. And the lesser difficulties were taken in stride and resolved along the way as RAND gradually laid the basis for a flourishing research enterprise.

THE CHARACTER OF THE EARLY RESEARCH EFFORT

After the letter contract of 2 March 1946, work on Project RAND began in earnest. The Douglas Company had taken a

---

[22] One is justified in speaking of the Air Force as a separate service even in these preunification days. Although independence from the Army did not officially come until 1947, it had been virtually assured since 1942 when Gen. Arnold sat as a coequal member with the Army and Navy representatives on the Joint Chiefs of Staff. See Paul Y. Hammond, *Organizing for Defense* (Princeton University Press, Princeton, 1961), pp. 113-131, 162, 167-168, 187.

number of preliminary steps toward activating the project even before the formal contract, but these were mainly of a housekeeping nature. Physical space was provided, elaborate steps were taken to seal off Project RAND administratively and physically from the normal Douglas operations, and the nucleus of a staff was assembled. Franklin R. Collbohm was put in charge temporarily, pending the appointment of a permanent director for the project.

At Air Force request, RAND undertook its first study in the spring of 1946 on the subject of space vehicles. Its first report came out on 2 May 1946 carrying the title "Preliminary Design of an Experimental World-Circling Spaceship." This study was typical of RAND studies during the first year in being oriented almost exclusively to "hardware" considerations. In the 324-page report, only scant attention was shown to what might be called the social-political-strategic implications of an earth-circling satellite.[23] There was only a brief consideration of cost estimates in sharp contrast to much later RAND systems-analysis work which has tended to be extremely sensitive to cost considerations in a broad sense. Throughout the year, Project RAND issued a number of follow-up studies on the earth-circling satellite as well as

---

[23] In retrospect, however, when one considers the furor created by the Soviet Sputnik launching in 1957, the few sentences in the report devoted to political implications seem prescient: "The achievement of the satellite craft by the United States would inflame the imagination of mankind, and would probably produce repercussions in the world comparable to the explosion of the atomic bomb . . . Since mastery of the elements is a reliable index of material progress, the nation which first makes significant achievements in space travel will be acknowledged as the world leader in both military and scientific techniques. To visualize the impact on the world, one can imagine the consternation and admiration that would be felt here if the U.S. were to discover suddenly that some other nation had already put up a successful satellite." Quoted by R. Cargill Hall, "Early U.S. Satellite Proposals," in *Congressional Record*, vol. 109, October 7, 1963, A6279.

technical memoranda on aerodynamics, heat-transfer problems of rockets, exotic fuels, range and design problems of
intercontinental bombers, and other related technical subjects.
Much of this early work was engineering science of a theoretical nature, along with various studies that resembled the
kind of research done by operations researchers in World
War II.

Progress was slow, however, and the results achieved during the initial phases of RAND's history proved disappointing
to some RAND supporters. Dr. Bowles was one who was particularly disappointed at the lack of early progress and he was
apprehensive about the way the project seemed to be developing. As early as 5 September 1946, he expressed his mis-
givings in a lengthy letter to General Arnold:

I am concerned over the Douglas project which, through your
vision, was initiated that October day in 1945 when we had a conference with Don Douglas at Hamilton Field . . . After heroic
effort and splendid perception and support on the part of LeMay,
we overcame the resistance of the Wright Field group and were able
to bring Air Plans and materiel people together on this project on
a congenial basis . . . I believe we set a precedent in recognizing that
we cannot do intelligent, long-range, strategic planning without
taking into consideration our scientific and technological resources
and their future development, nor can we give proper direction to
research and technological development without its leadership having
some concept of our strategic plans.

Now that all the obstacles within the military have been overcome,
it looks as though the real obstacles are within the Douglas Company
itself. I will review the reason for apprehension. In the first place,
when . . . the proposal [was first put] to me, it was on the basis
that Don Douglas, having had the satisfaction of building a great
organization and of making a contribution to aviation, was now
ready to go forward to bigger things. I was given to understand that
he wished to set up what, in effect, would be a foundation to under-
write outstanding research in aviation and that he was determined

51

this must be something entirely separate from the company's regular activities . . .

Since that time, I have got the distinct impression that this idealism has vanished from the picture and that we are working with Douglas on a strictly business basis, in which the Army Air Forces are underwriting any and all expenses of this project except what the Douglas Company may come by through taking over war assets.

Moreover, the project has been moving very, very slowly—so slowly, in fact, that the opposition in the Air Force is sitting by, if I may say so, gloating over the impending failure of what they consider an unrealistic and illy conceived enterprise. There is a growing "I-told-you-so" attitude. This is extremely disturbing to me, because I believe that in this Douglas project we have a type of activity that promises to give greater prestige to the Air Forces and greater credit than any other forms of contractual activity. As things stand, the stature you and I visualized for this undertaking has not even remotely been achieved.[24]

Bowles, meanwhile, had begun to grow out of touch with the project. Tied down in Washington, and learning of recent developments only through his capacity as Washington fence-mender for the project, Bowles was less and less able to make his influence felt in regard to the project's direction. By the time he left the War Department in August 1947, Bowles' influence over Project RAND's development had faded almost to nothing. He was deeply disappointed over the missed opportunity to create the joint industry-government planning mechanism for scientific development that he had envisaged since the early days of the war. But organizations, like children, seem to have a way of developing along different lines than their framers or parents intended. The idea of industrial sponsorship for such a venture was, in any event, to prove unworkable on any long-term basis for reasons that will become clear as the discussion proceeds.

[24] Bowles papers.

SUBCONTRACTING IN THE EARLY PERIOD

An interesting feature of RAND in the 1946-1947 period was that a large portion of the work was contracted out to other research organizations or industrial firms.[25] Much of this may have resulted from RAND's infant status and lack of sufficient staff. At any rate, for a time it looked as though the organization might blossom into a prime contractor with the actual research work being parceled out to a number of subcontractors. For several reasons, however, this practice was soon abandoned. One reason was that some of the work for which RAND contracted out was experimental in nature, and RAND management made a firm decision that experimental work or laboratory testing was not compatible with the RAND mission of background research on long-range problems.[26]

Another reason relates to one of the fundamental problems confronting Project RAND while it was attached to the Douglas Aircraft Company. Deep suspicion was aroused in other

[25] RAND's 1st Annual Report (RA-15035), March 1, 1947, showed that subcontracting costs for the first year of operation exceeded all other costs except labor and overhead, and were estimated at the time to become RAND's largest item of expenditure in years 1948 and 1949.

[26] One curious episode occurred early in 1947 that may have "burned some fingers" and had something to do with the decision not to engage in experimental work. *Business Week*, February 8, 1947, p. 5, reports: "Project RAND, the Army Air Forces' supersecret Buck Rogers department, is due to be canceled. Chief reason: Now that it's ceased to be supersecret, it sounds a little embarrassing. Project RAND is a contract with the Douglas Aircraft Company to maintain a staff of assorted experts who spend their time looking into fantastic or sideline ideas occurring to AAF brass. The experts were getting top-priority on anything they wanted . . . RAND came under a cloud when a requisition on Oak Ridge for a supply of radioactive isotopes resulted in a War Department investigation to find out who wanted the stuff. Project RAND didn't get the isotopes." It is perhaps worth reflecting on this episode in the context of general "laws" or "tendencies" of organizational behavior. New patterns of behavior and departures from existing policies seem to occur less by gradual accretive changes than by sudden dramatic incidents that force an organization to react sharply. For a further discussion of this point, see Chapter VIII.

aircraft companies by the anomaly of having an Air Force advisory group attached to a sister industrial firm and competitor for government R & D contracts. Other aircraft firms were understandably concerned about sharing their proprietary information with a research organization housed and nurtured by a leading competitor. They grew wary, therefore, of contractual links with Project RAND.

Finally, it became clear that extensive subcontracting only postponed the basic task of bringing researchers of varying backgrounds together under one roof to form a mutually stimulating and productive atmosphere. RAND management thus early recognized drawbacks to subcontracting, and by the late 1940's substantial subcontracting with other institutions was largely a thing of the past.

Today RAND generally makes a firm policy of not subcontracting-out research projects and not taking on assignments that it does not have the capability of doing within the organization. The device of the consultant, however, is still used to get the services for limited periods of outstanding individuals who are unable or unwilling to come to RAND on a more permanent basis. And on rare occasions RAND will resort to a subcontract with another research institution when it needs a research service performed that it cannot perform itself.[27]

The extent of RAND's early subcontracting raises the issue left unclear at the Hamilton Field meeting about how far Project RAND was to have broad industrial sponsorship. As previously seen, one idea held by some of RAND's framers was that the project should eventually broaden out into a planning

27 For example, RAND entered into a small subcontract with MIT in 1960 for a wind-tunnel test of a "ring-wing," an idea conceived by RAND engineer Roger Johnson for a concentric airplane to be used in some futurist airplane. A picture of Johnson studying a model of an exotic aircraft of this kind appears in *Life*, May 11, 1959, p. 101.

mechanism that would integrate the efforts of science and industry together with the military in planning for national defense. Although a certain amount of subcontracting may have been necessary when RAND was in the process of building up a staff, perhaps the full extent of the subcontracting actually done in those early years (particularly with other aircraft firms like Boeing, Northrop, North American) was in part inspired by the desire to secure the participation of other firms in the project at the working level. Participation at the top management level was supplied by the creation of the RAND Presidential Advisory Council in January 1947. The council was composed of the presidents of the Douglas, Boeing, Northrop, and North American aircraft companies, along with the Douglas Company's vice-president and chief engineer, Arthur Raymond, whose office was responsible for supervision of Project RAND. The council held a number of meetings throughout 1947 and early 1948, and generally tried to act as an over-all policy organ for Project RAND.

The arrangement did not work out very satisfactorily. Among other things, it ran up against the fundamental problem of Project RAND's intimate association with a manufacturing firm and the fear (even if groundless) of a conflict-of-interest developing between RAND's role as advisor to the Air Force and its connection with the Douglas Company. The concept of the Advisory Council never, in consequence, caught the enthusiasm of the other aircraft company presidents. As time went by, the Douglas Company itself grew increasingly disenchanted with the whole idea of RAND. Little real interest, therefore, was shown in the Advisory Council. The council met infrequently and attendance was sporadic. The last meeting of the council was held in February 1948 when the subject of RAND's separation from the Douglas Company was discussed; and there was unanimous agreement in the council on the desirability of the proposed separation.[28]

DOUGLAS SPONSORSHIP: ASSET OR LIABILITY?

Operating under the Douglas Company's fatherly support, though undoubtedly of great benefit to the infant organization in many ways, early proved to be a handicap to RAND as well. The advantages are clear enough. The relationship with Douglas was instrumental in getting the project off to a proper start. It remains unlikely that RAND could have achieved an auspicious beginning without the support of some established company like Douglas. On the negative side, minor irritants arose from time to time as administrative practices appropriate for an engineering firm conflicted with those suitable for a research organization. The matter of hours of work was an early instance. The Douglas Company employees worked regular hours and access to the plant was more difficult during the off-hours. Academic people accustomed to more flexibility in working hours found this irritating. The company also had a certain policy on coffee in the building because coffee could be spilled on drawing boards; this made it difficult for the RAND people to enjoy casual coffee breaks. Other examples abound. The net result was that the RAND people felt somewhat constrained under the Douglas sponsorship, and wanted the project to evolve toward a status that would assure greater flexibility of operation.

There were other more significant ways in which the Douglas association proved to be a handicap to Project

[28] Letter from Arthur E. Raymond to Edward L. Bowles, March 16, 1948. Apropos of the meeting Raymond wrote: "All members of the Council, including Clair Egtvedt [of Boeing Aircraft] were present and all expressed themselves as being in hearty agreement with the move."

RAND. One of these was the problem of attracting and keeping a research staff of a high quality. The natural antipathy of many academic people toward industry found expression in RAND's early years in difficulties of recruiting and holding some much-sought-after talent.[29] Consequently, although Project RAND was able to attract some highly competent researchers (many of whom were in the process of returning to private life after service in the government during the war) it was not able to attract enough outstanding individuals to build a research competence of the first order.[30] One need hardly add that in a research organization, especially one engaged in conceptual research, there is no substitute for a high-quality staff. This concern over staffing added to the growing pressure within RAND for an end to the Douglas association.

Also, and perhaps of primary importance, many RAND people felt that RAND would have difficulty achieving a reputation for truly objective research while it remained under Douglas sponsorship. A final break with Douglas seemed inevitable if the project was to move toward the status of a mature research organization.[31]

In May 1947, Project RAND moved physically out of the

[29] For the "flavor" of this problem in RAND's early years, see the comments of RAND old-timer J. D. Williams in R. D. Specht, "RAND—A Personal View of Its History," *Journal of the Operations Research Society of America* 8:826-827 (November-December 1960).

[30] A Raymond letter to Bowles comments in this regard: "While the project has some good men, it does not have as many as it needs and some that it does have are not as good as they should be. The situation in this regard has been improving of late and I am quite sure that it will be quite a bit easier to secure the services of more of the good people once the reorganization has been affected. The Douglas Company's connection has been a distinct handicap in this regard, although it has also made possible the recruitment of a number of very capable individuals in certain categories." Letter dated March 16, 1948 (Bowles papers).

[31] Some RAND framers seemed to have recognized from the beginning that the Douglas arrangement could only be of a provisional nature.

main Douglas building into rented quarters in downtown Santa Monica. This move both reflected the desire to grow away from the Douglas Company's fatherly support and in turn stimulated further thought about alternatives to Douglas sponsorship.

By the end of 1947, then, a series of pressures began to build up that were leading toward RAND's separation from the Douglas Company. The pressure came from several directions. First, for the reasons already suggested, RAND itself was growing increasingly restive about the Douglas relationship. Second, pressure was building up within Douglas to find some means for getting rid of Project RAND. The Douglas Company was entering on more difficult days, and there was some feeling within the company that Project RAND was a business disadvantage. The Air Force, many Douglas people believed, was forced to lean over backwards in the awarding of procurement and development contracts to avoid the impression of favoritism to Douglas. There was also considerable resentment within the company against the special ground rules of the Project RAND contract, and the feeling grew that nursing Project RAND along was more trouble than it was worth. Third, other aircraft firms grew more and more uneasy about the prospect of Air Force advisors operating under a chief competitor's auspices. RAND's move out of the Douglas main office into rented facilities only slightly allayed this suspicion for the project was still administratively attached to the Douglas Company and its personnel were still Douglas employees.

Finally, the Air Force itself felt uncomfortable about the RAND-Douglas connection. Air Force officials began to fear adverse criticism arising out of what might look like a possible conflict-of-interest situation. There was never any evidence that proprietary information obtained by RAND from other firms passed into the hands of the Douglas people. But the mere suspicion was enough to give pause to many Air Force officers who were reflecting on the matter. In government, avoiding the appearance of evil is often as important as avoiding the evil itself. The Air Force also had grown to suspect the company's good faith over certain difficulties in the administration of the contract. A particular sore point was the question of fixed overhead costs. Air Force contract officers objected that the company was making an excessive profit under the fixed overhead rate of the Project RAND contract.

In brief, the conditions were ripe for the break with Douglas and the separate establishment of RAND as an independent nonprofit research corporation. The story of how this was accomplished will be told in the next chapter. For present purposes, it will suffice to note that the foundations of a highly competent research organization had been laid by the time of the separation from Douglas. The nucleus of a high-quality staff was formed, and was to be added to in the years following the Douglas association. Franklin R. Collbohm was named permanent director for the project after there had been some early foundering due, on the one hand, to the company's initial reluctance to release some of its best people permanently to the project and, on the other, to the company's inability to attract an individual of sufficient stature from the outside to head the project.[32]

RAND management had learned much in those early years

[32] In mid-1947, the company decided to release Collbohm permanently to the project. Until then, he had been managing the project on a provisional basis pending the recruitment of a suitable director from the outside. Among the notable individuals outside the Douglas Company who were approached and who declined to serve as director of Project RAND were Lee A. DuBridge, President of the California Institute of Technology and Louis Ridenour, Dean of the Graduate College, University of Illinois.

about how to administer a research organization. A certain looseness in administration, for example, was highly desirable from the standpoint of promoting creativity in the research effort. "Over-administration" was found to be a grave danger. Also, the internal subdivisions in RAND were formed for the most part around professional-skill groupings rather than units corresponding to military operations such as strategic air, air defense, limited war, and so forth.[33] This had the advantage of facilitating recruitment, and of permitting a flexibility of approach that would not be obtainable if one group had a monopoly on studies in a given area. A fairly effective working relationship between the Project RAND people and the Air Force had also been developed. In sum, RAND was on the way to becoming a valuable aid to Air Force policy makers.

A clear lesson emerges from the Douglas sponsorship, however, which would seem to have relevance for any future policy research operation. On any long-term basis, the concept of an advisory group tied to a firm that also sells items of hardware to the government would appear simply unworkable. Objective technical advice and competition for hardware contracts are incompatible functions that cannot sensibly be combined in the same organization.

THE DEVELOPMENT OF THE SOCIAL SCIENCES AT RAND

We saw that RAND studies during 1946 were oriented predominantly toward mathematics and the physical and engineering sciences. This was to continue to be true, for the most part, throughout the next several years. During this same

---

[33] See Specht, "RAND—A Personal View of Its History," p. 830. For an extended discussion of RAND's internal organization, see Chapter V below. One can speculate that the nature of science may affect the general administrative theory that government agencies should be organized primarily by their general purpose, rather than by the nature of the skills involved in their operation.

period, however, the social sciences began to take root as a part of the RAND research effort. The idea that the social sciences should be part of a broad national-security research program was dimly present in the minds of some of RAND's framers and earliest employees. The original roster of RAND consultants had also included several social scientists.[34] As early as the winter of 1946, we begin to see some significant steps taken toward bringing social scientists into RAND on a permanent basis. The initiative for these moves rested to a large extent with John D. Williams, a mathematician and one of the early additions to the RAND staff.

After considerable debate within RAND about the need for social scientists and the feasibility of persuading the Air Force to accept the idea at that time, Williams received permission to go to Washington and try to "sell" the idea to General LeMay, Air Force Deputy Chief of Staff for Research and Development. A meeting with General LeMay took place in late 1946 where it was decided, after some discussion, that RAND should proceed with plans for integrating social scientists into its research effort.[35] The success of Williams' foray surprised many RAND people. Williams himself reflects, half-seriously and half-facetiously, that "they sent me to Washington to kill the idea so I'd stop

[34] 1st Annual Report (RA-15035), March 1, 1947.

[35] Four men were present at the meeting: Gen. LeMay, Gen. Thomas Power (later LeMay's successor as SAC Commander), RAND Vice President L. J. Henderson, Jr., and Williams. LeMay originally scoffed at the idea of adding social scientists, and stressed the need to spend the taxpayer's money for operational weapons. Gradually, however, Williams persuaded him that it would be a good investment to spend a small amount of money for research on how to use the weapons in our arsenal. At the end of the meeting Williams asked if he rightly understood LeMay's decision as authorizing RAND to proceed cautiously and experimentally by adding a few qualified social scientists to the research staff. Gen. LeMay replied, "No, no, that's not it. Let's do it up right. If we're going to do this, do it on a meaningful scale." Interviews with John D. Williams.

61

pestering people."[36] Again, the Air Force deserves credit for its enlightened administration of Project RAND. In retrospect, it appears remarkable that the Air Force was receptive to the social sciences at a time when the need for civilian assistance in the physical sciences had only recently been recognized.

Social science was much longer in coming to certain of the other nonprofit research institutions. It was not until 1959, for example, that the Navy's Operations Evaluation Group (OEG)[37] added three economists to its staff of physical scientists and mathematical operations researchers. Since the McNamara "revolution" in Pentagon management practices, the Navy has embarked on a program to augment substantially the social science component of its contract analytic facilities.[38] Similarly, in 1962 the Institute for Defense Analyses (IDA) began to organize an economics division (although it had a small international studies division composed mostly of political scientists for several years prior to that).[39] The Army's Research Analysis Corporation (RAC) seems to most closely parallel RAND in having had social scientists on its staff from an early date. By 1953, over 40 percent of its professional staff were from economics and human arts and sciences.[40]

[36] Interview, Santa Monica, California, December 11, 1961.
[37] Since 1962 part of the Center for Naval Analysis.
[38] Interview with Dr. Frank Bothwell, Director, Center for Naval Analysis, in Cambridge, Mass., October 25, 1962.
[39] Interview with Dr. Stephen Enke, Assistant to the IDA President, Secretary of the Research Committee, and head of the IDA Economics Studies Division, Washington, D.C., April 2, 1963.
[40] Dr. Lynn H. Rumbaugh, Research Analysis Corporation, "A Look at U.S. Army Operations—Past and Present," unpublished paper, June 1963, p. 10. The combined fraction of the social scientists, broadly defined, still numbers 40 percent of RAC's staff. Significantly, however, within the 40 percent the percentage of economists has gone up markedly while the humanists have declined sharply in percentage.

Subsequent to Williams' Washington visit, a conference of social scientists was arranged in New York to explain Project RAND to the social science community and hopefully to interest some able people in joining the RAND staff. Warren Weaver, President of the Rockefeller Foundation, participated in the conference and gave the keynote address. He also generally assisted RAND in organizing the conference, which was held in September 1947 and was helpful in recruiting some of the men who played leading roles in the development of the social sciences at RAND. Hans Speier and Charles J. Hitch were among the social scientists present at the New York conference. These men became, respectively, head of the Social Science and Economics Divisions.

Interestingly, the social sciences developed along generally two distinct lines at RAND. Separate Social Science and Economics Divisions were established, and these evolved in different ways. The Economics Division became one of the most prominent research units in RAND, and through the years did much to give RAND work its distinctive character. That is, the Economics Division helped impart an awareness of cost considerations in the broadest sense and their importance in guiding choice under uncertainty—a way of looking at military problems that has come to figure prominently in the RAND research effort. It is not accidental that the economists have come to form between 15 and 20 percent of RAND's research staff.[41]

[41] As of September 30, 1965, there were 78 economists out of 501 professionals on the RAND research staff, or 15.6 percent of the total. The percentage has remained fairly constant over the 1960-1965 period, sometimes increasing slightly but never dropping below 15 percent. If one considers different kinds of engineers as separate categories for purposes of the tabulation, the economists are the largest single professional skill group on the RAND research staff. However, when all the different engineers (aeronautical, chemical, electronic, mechanical, and other) are considered as one group, they outnumber the economists 145 to 78. The next largest skill groups after the economists

The social scientists, on the other hand, did not enjoy the spectacular growth of the economists and for a long time remained on the fringes of RAND's major research activities. This is true in both a literal and a metaphorical sense. For the geographical isolation of most of the social scientists from the main RAND headquarters contributed importantly to a separate identity and feeling of spiritual isolation from the rest of RAND that seemed to characterize the Social Science Division for a number of years after its creation. It was not until 1956 that the bulk of the social scientists finally moved from RAND's Washington office to the main headquarters in Santa Monica, California. There seems little doubt that the move has improved interdisciplinary communication and generally worked to the benefit of both the RAND social scientists and their colleagues in other disciplines. Social scientists have come to figure prominently in most of RAND's important systems-analysis work and in general to enjoy high prestige among their RAND colleagues. Political scientists make up the largest skill group in the Social Science Department, but such fields as psychology, history, philosophy, law and anthropology are also represented.

It is too much to say that the social sciences were a part of the RAND concept from the beginning, but by the same token it would be inaccurate to say that the social sciences were simply added as an afterthought. As early as the Special Bombardment Project of 1944, Raymond and Collbohm were groping toward a kind of cost-effectiveness model as an aid to decision on strategic-weapons systems. More likely the social sciences were thought of early in RAND's history but there was no clear idea of how they would be used or when

---

and engineers are mathematicians, physicists, cost analysts, and computer programmers. (Material obtained from the RAND Personnel Office.)

an appropriate staff could be assembled. The success in persuading the Air Force that social scientists were necessary intensified thinking about the social science role, and accelerated the time schedule on which they would be brought into the organization. The addition of the social sciences was necessary sooner or later because the broad systems-analysis work that RAND sought to do could not be done effectively without social science skills. While the bulk of RAND's work consisted of simple operations-research problems of the World War II variety, the social sciences were not so necessary. But as the problems RAND dealt with became more complex the social science input became vital.

► III ◄

THE FORMATION OF THE
RAND CORPORATION

A non-profit corporation . . . formed . . . to further
and promote scientific, educational, and charitable
purposes, all for the public welfare and security of
the United States.

—*RAND Articles of Incorporation, 14 May 1948*

In late 1947, the pressures leading toward RAND's separation from the Douglas Aircraft Company began to build up.
From 10 September 1947 serious consideration began to be
given to alternative arrangements. On that date, Arthur E.
Raymond wrote to Edward L. Bowles stating the Douglas
Company's intention to give RAND some sort of autonomous
status:

. . . RAND has good momentum now and enjoys fairly healthy
prestige in most quarters in Washington. It has, I believe, grown
up and the time is fast approaching when it will not need the same
degree of fatherly support which the Douglas Company has thus far
provided. As a matter of fact, it is going to be desirable, both from
the standpoint of RAND and the standpoint of Douglas, that a state
of almost, if not complete, autonomy be established as rapidly as
this can practically be done without injury. The end point should
be one in which the relationship of Douglas men to the working
[RAND people] is exactly the same as representatives from other
companies.[1]

[1] Bowles papers.

66

That the separation was desired by all parties, however, does not mean that it occurred without difficulty. In fact, the difficulties involved were considerable and it was not immediately obvious what sort of organizational framework should replace the existing one.

RAND SEPARATES FROM DOUGLAS

A first difficulty concerned the legality of transferring the Air Force contract to another sponsoring organization. When Donald Douglas, Sr., was first approached in late 1947 about the possibility of separating RAND from the Douglas Company, he was skeptical about the prospects of accomplishing this under existing government-contract law. The matter was accordingly referred to the Douglas legal advisor, Frederick Hines, who came to the conclusion that it was impossible legally to contemplate such a move while moneys under the contract were left unexpended.[2] For a time the matter was simply dropped, and RAND's director was instructed not to proceed at present to discuss the matter with the Air Force.

Then Franklin R. Collbohm, Director of Project RAND, decided to call on H. Rowan Gaither, a friend and San Francisco lawyer, for legal advice and suggestions as to a way out of the impasse. Gaither played a vital role in the succeeding course of events that led to RAND's establishment as an independent nonprofit entity. And, as a trustee and principal advisor, Gaither was also to play a leading role in subsequent RAND affairs until his death in 1961. He gave the legal opinion that the difficulties would not be insuperable in getting the contract transferred from Douglas to some other appropriate body. He and Collbohm considered various alternatives to the Douglas arrangement and eventually concluded that the formation of an independent nonprofit corpo-

[2] At that time, Project Rand had spent just under half of the original $10,000,000 set aside for the project.

ration chartered under the laws of California would be the most desirable solution. As in the initial decision to locate in Douglas Aircraft two years before, there seemed to be serious objections against either an in-house government operation or affiliation with a university. The two men rejected other industrial sponsorship because of the conflict-of-interest question and the problems of securing effective access to other firms' proprietary information that arose under Douglas sponsorship. The not-for-profit corporate device was finally chosen as the most desirable remaining alternative because it was relatively easy to establish and it would afford maximum flexibility of operation.[3]

The matter was brought up again with Douglas, and he agreed to have the RAND people broach the issue with the Air Force. The Air Force's reaction is expressed in a letter dated 10 February 1948 from the Air Force Chief of Staff, General Spaatz, to Donald Douglas. The letter noted the evident advantages of the move to all parties concerned, and agreed that the move was now feasible whereas it had not appeared to be so at the inception of the project. Promising the full support of the Air Force, General Spaatz asserted that the Air Force could legally consent to the transfer of the contract to the new corporation. However, he set several conditions that had to be met before the consent would be given:

> This consent we are willing to give when we are satisfied that the new corporation is *in existence and is capable of discharging the contract obligations as effectively as the Douglas Company.*[4]

This point is worth emphasizing, for it helps throw light

[3] For a general discussion of the not-for-profit corporation and the other forms that a "philanthropic foundation" may assume, see F. Emerson Andrews, *Philanthropic Foundations* (Russell Sage Foundation, New York, 1956).

[4] Letter obtained from RAND files. Gen. Spaatz to Donald Douglas, February 10, 1948. Italics mine.

on the broad question of how RAND has been able to avoid a completely dependent status *vis-à-vis* the Air Force. The corporation was organized and set up independently of the Air Force—in short, came into existence *first*—and entered into the contractual relationship with the Air Force only *after* it was already established. This has not always been the case with other nonprofit corporations operating under the general sponsorship of one of the services; some of the other organizations have been the direct product of service initiative. They have been, in effect, conceived, born, and raised by the service itself. This factor may be of some consequence in influencing the degree of independence that the organization will have from direct sponsor control.

An illustration of the contrast between RAND and certain other defense nonprofit corporations in this respect may be found in the provisions relating to the disposal of the corporate assets upon dissolution. RAND's articles of incorporation provide that upon the dissolution of the corporation all assets will be distributed at the direction of the Ford Foundation for scientific, educational, and charitable purposes.[5] In contrast, the Aerospace charter includes, at Air Force insistence, a provision stating that the assets of the corporation would devolve upon the government and would be disposed of by the Air Force in the event of dissolution. Similarly, MITRE's charter provides for reversion of assets as directed by the President of the United States. The contract between the Army and the Research Analysis Corporation (RAC) also includes a similar provision which declares that the title to all property acquired by the corporation shall pass to and vest in the U.S. Government.[6]

[5] If the Ford Foundation has ceased to exist, the articles of incorporation provide that the Superior Court of California will dispose of all RAND assets.
[6] Contract document, Section 22, paragraph b, p. 31 (Armed Services Pro-

With the doubts as to the legality of the contract transfer removed by General Spaatz' letter, the way was cleared for efforts to begin on the actual formation of the new corporation. An important first step concerned the financing of the new corporation. Collbohm and Gaither decided it should have a $1,000,000 working capital availability to assure a separate existence and to convince the Air Force that it was a going concern. As a first step, RAND management approached a number of local banks to see if bank loans could be arranged. Initial efforts in this direction were unsuccessful. Subsequently, however, through the influence of the Gaither family several San Francisco banks agreed to back the project. They agreed to extend lines of credit of $600,000 provided RAND could first demonstrate that it had $400,000 in other working capital.

RAND officials then turned to the next most obvious source for more permanent financial assistance—the foundations. Here, again, initial efforts were unsuccessful. The several older foundations, RAND management discovered, had long-standing policies governing their operations that prevented them from considering capital aid to the new RAND Corporation. Around this time, however, the Ford Foundation was in the process of a reorganization which was to transform the foundation from a small philanthropy, essentially dominated by the Ford family, to the largest foundation in the country, managed largely apart from the control or influence of the Ford family. Vast new sums of money were in prospect, and the Ford family had indicated its intention to enlarge the Board of Trustees to appoint men of broad experience to the new positions.[7] As an essentially new enter-

curement Regulation 13-503, May 1961). Copy of contract obtained from Army Research Office, Arlington Hall Station, Arlington, Virginia.

[7] See Andrews, *Philanthropic Foundations*, pp. 26-29.

prise in large-scale philanthropy, the Ford Foundation had not yet crystallized policies governing its operations. Hence its freedom of action was not limited with respect to providing capital support to an organization like RAND. Moreover, H. Rowan Gaither as well as Raymond, Collbohm and Henderson had personal contacts with important people associated with the foundation. For these reasons the Ford Foundation was considered an ideal prospect for obtaining possible capital assistance. Meanwhile, the unavailability of credit elsewhere by this time had lent a note of urgency to the efforts of RAND management to find financial backing for the new corporation. Accordingly, plans were made to approach various persons associated with Ford.

One such approach that reportedly took place in the summer of 1948 has an almost storybook quality.[8] Three RAND people, learning that Dr. Karl Compton of MIT, a Ford Foundation trustee, would be traveling on a certain train from New York to Boston, arranged to occupy seats in the same compartment with him. On the trip to Boston, they managed to impress Compton with the new corporation and by the time the train steamed into Boston the RAND Corporation had a backer. A second Ford Foundation trustee, Donald David, Dean of the Harvard School of Business Administration, was approached shortly thereafter in Cambridge. David happened to be a close friend of RAND executive L. J. Henderson. He, too, was impressed with the plans for the new corporation, although reportedly somewhat less so than Compton. In any event, through the help of informal

8 Although memories were hazy, F. R. Collbohm and L. J. Henderson in separate interviews independently recalled this incident. I conclude that there is some substance to the story, although it is quite possible that in retrospect the incident appears more dramatic than it actually was. Some of RAND's framers had known Compton from the war and would have had little difficulty arranging a less dramatic meeting.

contacts like these, a subsequent meeting with Henry Ford II was arranged in Detroit. At this meeting Ford agreed to make RAND a $100,000 interest-free loan and to guarantee $300,000 in bank credit. The $600,000 from the San Francisco banks was then available, and a major hurdle was cleared. As it turned out, however, RAND decided it did not need the full $600,000 and actually borrowed only $150,000 from the Wells Fargo Bank of San Francisco.

But still another problem had to be faced before the RAND management could demonstrate to the Air Force that the new corporation was "in existence" and "capable of discharging the contract obligations as effectively as the Douglas Company." An outstanding Board of Trustees was needed to lend prestige to the fledgling corporation and to convince the Air Force that the new organization would have competent direction and the backing of the scientific and industrial communities. RAND was as yet not very widely known outside of government and aircraft-industry circles and by no means fully tried and tested as a research organization. It is not surprising, then, that eminent individuals might hesitate before becoming associated with the new corporation as trustees. Some of the persons first approached did, in fact, decline to serve. Nonetheless, RAND director Collbohm decided not to lower his sights and persevered until a group of prominent individuals was found.[9] This original decision to fill the Board with men of high standing has paid off handsomely for RAND over the years. Not only have there been dividends in the form of prestige and respect accorded the organization, but on several occasions the benefits have been

---

[9] The original members of RAND's Board of Trustees were: H. R. Gaither, Jr.; F. R. Collbohm; L. J. Henderson, Jr.; Clyde Williams; Philip M. Morse; Lee A. DuBridge; Charles Dollard; Frederick F. Stephan; John A. Hutcheson; Alfred Loomis; and George D. Stoddard.

more immediate and tangible. RAND trustees have helped blunt attacks on RAND by virtue of the high esteem critics have felt toward them personally. It has not been unknown in RAND's history for a trustee to help smooth out contract difficulties with a client as well.

Once the Board of Trustees was formed, the last major obstacle was surmounted and arrangements were completed with the Air Force for the transfer of the Project RAND contract from the Douglas Company to the newly founded RAND Corporation. The Douglas Company initially agreed to the release of all facilities (books, office equipment, and so on) connected with Project RAND to the new corporation. This, however, was vetoed by the Douglas Board of Trustees and the facilities were eventually sold to the new corporation at their book value of $12,000. The Douglas Company permanently released a number of key Project RAND people to the new corporation.[10] Provision was made for the extension of the unexpended portion of the original $10,000,000 contract sum to the new corporation, along with a small additional appropriation for the fiscal year 1949.[11] On 4 November 1948, the Air Force officially announced that a contract was established with the new organization for the broad

[10] Again the accidental assisted RAND. It happened that at this time there was an internal shake-up in Douglas management, and releasing some people to RAND provided a convenient way for the company to ease new men into certain important positions. Confidential interview.

[11] For a listing of the annual appropriations for Project RAND from Fiscal Year 1949 through an estimated amount for Fiscal Year 1960, see *Department of Defense Appropriations for 1961*, Hearings before the Subcommittee of the Committee on Appropriations, House of Representatives, 86th Cong., 2d Sess., part 7, p. 184. A similar and more up-to-date listing (except that the figures refer to total government support per annum for the RAND Corporation) may be found in *Systems Development and Management*, Hearings before a subcommittee of the Committee on Government Operations, House of Representatives, 87th Cong., 2nd Sess. (Holifield Subcommittee Hearings), part 3, p. 1100.

scientific study of methods to improve national security.[12] The date of the initial chartering of the corporation and the adoption of the by-laws was 14 May 1948. The purpose of the RAND Corporation, as stated in the articles of incorporation, was "to further and promote scientific, educational, and charitable purposes, all for the public welfare and security of the United States of America."[13]

THE LOCATION FACTOR

A last interesting feature in RAND's founding was the location of the new corporation. The new corporation could have been chartered and/or located in any one of a number of states or the District of Columbia. That it was located in California can be attributed in part to simple administrative convenience but also to the conviction that long-range thinking and research were difficult in the hectic pace of Washington life. The wisdom of getting out of the Washington vicinity seemed to have been amply demonstrated by the two years experience since 1946; and RAND management decided therefore to incorporate the new organization under California law and to continue residence in Santa Monica.[14]

The evidence seems to suggest that there was logic to this decision. The research institution located in or near Washington seems more likely to become absorbed in day-to-day

[12] Press release obtained from the RAND Corporation files. A copy of the press release appears in the *New York Times*, November 5, 1948.

[13] Article 2, paragraph a, Articles of Incorporation. The RAND Articles of Incorporation are reprinted in part 3 of the Holifield Subcommittee Hearings. The RAND incorporation document follows closely the model Not-For-Profit Corporation Law of the American Bar Association; the corporate purposes are broadly worded so as to afford a maximum flexibility in operation.

[14] In November 1948, however, a small branch office was established in Washington, D.C. This office housed mainly administrative and liaison staff, but as previously noted the bulk of RAND's social scientists were also located in the Washington office until 1956. Until 1963 RAND also had a small liaison office at Wright Field in Ohio.

operational problems and short-range projects than the institution located some distance from Washington. To illustrate this point, let us contrast briefly the experience of Analytic Services, Inc. (ANSER) with that of RAND. ANSER, it is true, was created to fulfill a markedly different, more short-range research function than RAND. The choice of a location near Washington was in this sense understandable and appropriate. Even acknowledging this fact, however, one must conclude that ANSER's location hardly more than a mile from the Pentagon has had a marked impact on the group's operations. It is doubtful whether ANSER could do much significant longe-range research even if it wanted to or had been established for this purpose. The Fairfax County location has led to ANSER being called on for a large number of short-run or "crash" projects, on some occasions even speech-writing and other normal staff activities. In effect, ANSER is a kind of auxiliary staff, a catch-all for short-run requests from the Director of Development Planning, Headquarters USAF.

The fact that the Institute for Defense Analyses (IDA) has most of its research staff in the Washington area is perhaps also of some significance. It may help explain why a substantial part of the IDA research effort is generated by a system of "task orders" (that is, written requests for studies) coming from the sponsoring government agency— a practice which contrasts rather sharply with the way in which Project RAND studies are typically initiated.[15] Also, the issue of how much "independence" the advisory institu-

15 See Chapter V. The IDA "task order" system apparently was somewhat modified in 1963, however, with the result that IDA now resembles RAND more closely in the way studies are initiated. It should also be noted that much of RAND's non-Project RAND work resembles the "task order" system of initiating studies.

tion shall enjoy is not wholly unrelated to the question of geographical location.[16]

With today's ease of communication and transportation, RAND researchers are finding that even geographic separation from Washington is no guarantee that they will escape the pressure of today's crises and operational problems. Unusual self-discipline is required to prevent a stream of telephone calls and visitors from severely impinging on research time.

On the other hand, the advantages are not all on the side of a location away from Washington. Physical separation poses some basic difficulties in communicating the research results to the appropriate decision makers in a timely and effective fashion. The advisory organization located in Washington, although it may be more subject to harassment and day-to-day pressures, at the same time has the advantage of close contact with operating officials and a sympathetic hearing for the results of work that has been carried out with the operator's knowledge and support. By the time a study emerges from ANSER or WSEG, for example, it is usually "half-sold" because of the intimate contact the researchers have had with operating officials throughout the study's formative stages.[17] RAND's physical separation from sponsors has, in short, not been an unmixed blessing.

A FRAMEWORK FOR RAND–AIR FORCE RELATIONS: THE AIR FORCE POLICY STATEMENT OF JULY 21, 1948

An important Air Force policy statement, "Air Force Policy for the Conduct of Project RAND," was issued during the transition period when RAND was separating from Douglas

---

[16] See Chapter VII.
[17] For a more complete discussion of the communication problem, see Chapter VII.

and incorporating as an independent research organization.[18] This statement is worth examining in some detail because it provides the basic setting for the broad pattern of Air Force-RAND relations that developed over the next decade. In part the statement only codified existing practices, but it also laid down some important general guidelines that helped shape future RAND–Air Force relations. It began by describing the initiation of Project RAND in March 1946, noting the original statement of work formulated at the LeMay meeting, and endorsing the pending separation of RAND from the Douglas Company. Then it went on to make the significant pronouncement that "RAND is a background research project—not a development project." Investigations undertaken by RAND were normally not to be carried beyond the point of "obtaining the information necessary to prove the validity of theoretical assumptions made" or the background information necessary to "prepare a type specification which can be used by the Air Force in negotiating a product development program." These somewhat ambiguous statements did not entirely clarify how far RAND might go toward leaving the stage of purely "paper" research or getting involved in the granting of contracts for military hardware, but they nonetheless provided the underpinning for what has been and continues to be a firm RAND policy: RAND will not engage in experimental or developmental research or comment directly on industry proposals for particular weapons-development programs.

Several years later, an Air Force study was to complain that RAND had isolated itself to some extent from real weapons by not getting involved in evaluations and by its refusal to participate in analyses which might lead to the

[18] Air Force letter 80-10, Department of the Air Force, July 21, 1948. Issued over the signature of Air Force Chief of Staff, Hoyt S. Vandenburg.

granting of a contract in competition.[19] But as the success of a large part of RAND's research work depends on access to information from a variety of industrial contractors, it is difficult to see how RAND could avoid a policy of not becoming directly involved in analyses of specific weapons-development programs. There is a fine line to be drawn, however, between background technical-feasibility studies and direct comments on a specific industrial-development program. In actuality, RAND does become involved on occasion in commenting on specific developmental programs but usually these activities are incidental to the main RAND research effort. A brief glance at some RAND responses to Air Force requests for informal assistance for the first half of 1960 points this up well. RAND gave such informal assistance to the Air Force as "Participation as Observers in Evaluation of Contractor BAMBI Proposals," "Evaluation of Industry Proposals of Future Air Transport Requirements," "Comments on Boeing Friendly Interference Study," "Atlas and Titan Missile Costs," and "Minuteman System Cost Estimates."[20] RAND can be particularly useful, in such informal assistance, in helping the Air Force to evaluate impartially the cost estimates of contractors' development proposals.

Another interesting point contained in the Air Force Policy Statement was the affirmation that Project RAND will "continue to have maximum freedom for planning its work schedules and research program." Thus the early enlightened policy of Generals Arnold and LeMay in administering Project RAND became firmly enshrined in official doctrine. This

[19] Internal Air Force study, n.d. Obtained from the Bowles files. From the content of the study, I infer that the date was about 1952.

[20] Internal RAND Document, Listing of Responses to Air Force Requests for Informal Assistance, 1960.

gave important impetus to the new corporation's efforts to regularize a pattern of working relationships with the Air Force that would permit both wide latitude in managing its own affairs and close contact with all parts of the Air Force. Of particular significance in this kind of relationship is the freedom the research group has to initiate studies and to define the problem under investigation. Throughout its history, RAND has enjoyed the freedom to refrain from doing studies it does not feel equipped to perform or thinks incompatible with the RAND mission. Establishing this pattern of favorable working relationships early has contributed in no small part to RAND's freedom and relative independence from sponsor domination over the years. Momentum and tradition tend to be built up around a given way of doing things, and innovators face the always difficult tasks of overcoming inertia and altering established procedures. Hence, those who have sought to curtail RAND's independence have run up against traditional patterns and have found strong forces working against them. The task of those within the Air Force who have guarded RAND's freedoms has in this sense been made correspondingly easier. In part, then, the answer to the question of how RAND has been able to enjoy through the years a substantial measure of freedom and independence is deceptively simple: it got started that way. Through a happy combination of chance, good fortune, and foresight on the part of a small group of people, RAND got started on a firm footing which proved to be a great asset in years to come.

Here, again, it is important to note the auspicious timing that aided RAND's development. At the time of the incorporation and the issuance of the Air Force Policy Statement, the Air Force was still in a position of undisputed leadership in the national military establishment. She had little to fear from competing sister services, and her sole possession of nuclear weapons made her more powerful than any potential foreign enemy. The Air Force could, in a sense, afford to be magnanimous and permit a research group to think about long-range problems at Air Force expense with a minimum of control and direction. The Air Force had little cause to worry about RAND's relationship with other parts of the national defense establishment or little need to draw RAND into consideration of immediate operational problems. Now in an era of strong competition among the services for shares in the defense budget and an unprecedented threat from a foreign enemy, it is perhaps doubtful whether any similar long-range research organization could be established at the service level with the same degree of independence and freedom that RAND has enjoyed. There are stronger pressures to watch the studies and pronouncements of such an organization, insofar as they deal with strategic or inter-service matters, out of regard for the service's position within the defense establishment. And instead of a long-range perspective, the pressures are strong to draw the research group into pressing short-run problems. RAND has not escaped the impact of these pressures, and in recent years has had to struggle against the tendency to be drawn more closely into current policy issues and pressing operational problems. More will be said on this in later chapters. For present purposes, let us consider several related points contained in the policy statement of 21 July 1948.

The use of Project RAND, the policy statement declared, to accomplish specific "crash program" staff work will be minimized. RAND was neither conceived nor staffed, it was asserted, as an organization to provide quick answers for current problems. The implications of this part of the policy statement are clear: it meant that RAND was to be spared much unnecessary interference with its research program.

To help carry out this policy, the statement provided that all requests for studies by RAND were to be channeled through a central monitoring office within the Air Force for Project RAND affairs.[21] In practice, the system has not worked in quite the neat and orderly fashion envisaged in the policy statement. Studies are often generated through informal contacts with parts of the Air Force other than the central monitoring office. But the system has provided for a measure of orderliness in the administration of the large number of requests and suggestions for studies that come from various parts of the Air Force. And it has also spared RAND the unhappy prospect of an internal power struggle with various parts of the Air Staff or particular commands trying to capture the organization for their own exclusive use. Such a situation might well have occurred in the absence of a high-level referee for RAND affairs within the Air Force.

However, the policy statement has not freed RAND from *all* requests for short-run staff assistance or the "quick and dirty" research job. The wording of the policy statement on this point is itself instructive: the use of Project RAND for such purposes is to be "minimized" rather than "forbidden." RAND does comply, in fact, with a number of such short-run requests. This assistance can take many forms, such as briefings, simply supplying information already at hand, comments on a broad study in progress, or the loan of staff temporarily to an Air Force unit. The amount of effort involved, in terms of research man-hours, can be very small or fairly considerable. In each instance, it is a question of judgment by RAND management and by the individual researcher to decide whether complying with the request will provide a

[21] The Deputy Chief of Staff for Research and Development and a special Project RAND Liaison Office set up within that office.

useful service to the sponsor at an allowable cost in research time and effort. During the calendar year 1964, RAND fulfilled 285 requests for short-run staff assistance or "quick response" studies.[22] The most serious danger of this type of activity is that it represents a heavy burden on key people who are the vital creative cogs in the research effort.

In another interesting point, the policy statement spelled out the official Air Force view of outside contractual work by RAND. The new RAND Corporation "will be free to undertake supplementary work for agencies other than the Air Force, or jointly for the Air Force and other agencies, provided suitable financial arrangements are made and the work does not interfere with Air Force studies." This was to supply the justification for subsequent RAND moves, beginning with the Atomic Energy Commission in 1949, to diversify its base of support by obtaining research contracts with additional government agencies. The Air Force could hardly have foreseen how far RAND was to move in the direction of obtaining additional sponsorship. Had it foreseen this development, the Air Force might have had some second thoughts about laying down such a permissive and unqualified policy on outside contract work by RAND. Now that RAND has moved toward additional sponsorship at the Office of the Secretary of Defense level, many in the Air Force are beginning to feel uneasy and to wonder whether such sponsorship is compatible with RAND's confidential advisor relationship with the Air Force.[23]

---

[22] Internal RAND materials. The 285 figure represents roughly a 40 percent increase over a comparable figure for 1961. The data are not entirely reliable, however, because the definition of "quick response" assistance is rather ambiguous. But the general inference one can draw is probably accurate: in recent years RAND has increased considerably the "quick response" component of the research effort.

[23] It would be a cardinal sin, of course, for RAND to contract with one of

Finally, the policy statement is noteworthy for its recognition of the importance of continuity and stability to the success of the RAND research effort. The project would have to be, the statement announced, "unusually stable to be effective." Hence, adequate fiscal support was to be provided to insure the continuity of the project. Such continuity and stability was particularly required because of "the extremely high caliber of personnel required to conduct this background research." Despite occasional vicissitudes in funding, there has been over the years a high degree of continuity in the level of Air Force support for Project RAND.[24]

The fee-income has been important in this context in permitting the accumulation of working capital to tide the corporation over periods of uncertainty in the scheduled availability of government funds. (Both before and after its separation from Douglas, RAND has operated on a cost-plus-fixed-fee contract with a fee of 6 percent.) The fee has also been important in helping maintain continuity and excellence in RAND staffing. The availability of fee-income for special self-sponsored research efforts has provided needed respites for RAND researchers from the demands of classi-

---

the other services. An internal Air Force study (cited in note 19 above) once declared in this regard: "The lawyer-client relationship of RAND to the Air Force places upon RAND certain restrictions. It is inevitable that the three Departments of the National Military Establishment will compete for budgets, facilities, and military responsibilities. As a result, it is inappropriate for RAND to 'represent' more than one of the services."

The likelihood of one service's advisory organization contracting with another service for research work seems extremely remote at present. However, the Hudson Institute, set up in 1961 by former RAND physicist Herman Kahn, performed a modest level of research work for several services during the first year of its operation. Interviews with Herman Kahn, Donald Brennan, and Max Singer of the Hudson Institute.

[24] See note 11 above. The importance of continuity in funding for research projects, and the deleterious consequences of abrupt fluctuations in support levels, are discussed in J. Stefan Dupré and Sanford A. Lakoff, *Science and the Nation* (Prentice-Hall, Englewood Cliffs, N.J., 1962), pp. 38-40.

fied national-security research. The Bell Report has strongly endorsed the practice of paying fees to not-for-profit organizations (other than universities) on the grounds that "such allowances provide some degree of operational stability and flexibility to organizations which otherwise would be very tightly bound to the precise limitations of cost financing of specific tasks" and because "most not-for-profit organizations must conduct some independent, self-initiated research if they are to obtain and hold highly competent scientists and engineers."[25]

During times of some special need or fluctuation in Air Force support levels, RAND has been fortunate in obtaining certain assistance from other bodies. The Ford Foundation, for example, in 1950 raised its initial $100,000 interest-free loan to $1,000,000 to permit construction of a headquarters building and to facilitate the expansion of a program of RAND-sponsored research. Several years later, the Ford Foundation converted the loan to an outright grant which made it possible for RAND to continue uninterruptedly its program of self-sponsored research. Additional sums have also been given recently by the Ford Foundation in the form of grants for research on urban transportation problems and for a study of the possible applications of systems analysis to educational problems.[26] The National Science Foundation, the Carnegie Corporation, and the Rockefeller Foundation have also supported particular research projects.

[25] Bell Report, p. 41, reprinted in Holifield Subcommittee Hearings, part 1, p. 237.

[26] This latter assistance resulted in the publication of a RAND research memorandum by Joseph A. Kershaw and Roland N. McKean, *Systems Analysis and Education*, RM-2473 (October 1959). A related study, financed half by the Ford Foundation and half by RAND's own funds, was published as a book by Kershaw and McKean, *Teacher Shortages and Salary Schedules* (McGraw-Hill, New York, 1962).

RAND DIVERSIFIES: THE ATOMIC ENERGY COMMISSION CONTRACT

The Air Force Policy Statement of 21 July 1948, as seen above, opened the way for a possible broadening of RAND's sponsorship to agencies other than the Air Force. The first instance of RAND actually obtaining additional sponsorship was not long in coming. In 1949, all the services felt the impact of a congressional economy wave. In consequence, the Air Force informed RAND in May 1949 that because of budget limitations and increasing requirements for a force-in-being it would have to conform to a lower annual expenditure rate than the rate to which "earlier informal advice" had urged the organization to expand "as rapidly as was consistent with your ability."[27] RAND would have to operate within a ceiling of $3,000,000 instead of $4,000,000 as anticipated earlier, and to this end the provision for the coming year would be tailored down to achieve a leveling off at the $3,000,000 annual rate by fiscal year 1951. The Air Force echoed the policy announced the previous year in indicating that it would have no objection to RAND's contracting with other agencies for analytic research similar to that done under Project RAND. But "in the event such other work is contemplated," RAND was told, the Air Force "would appreciate the opportunity to consult in advance with the RAND officials on the prospective work."[28]

Faced with the unhappy prospect of scaling down the scope of its activities and losing a part of the research staff, RAND management determined that the most desirable course of action would be to seek additional sponsorship elsewhere.

[27] Letter from Brig. Gen. (now Lt. Gen., ret.) Donald L. Putt to RAND President Franklin R. Collbohm. In the following discussion of RAND's contract with the Atomic Energy Commission, I am indebted to a short internal RAND monograph by Vaughn Bornet which recounts certain phases of RAND's history.
[28] Putt letter.

RAND President Collbohm accordingly informed the Air Force in August 1949 that "we contemplate tapping other sources of sponsorship to provide what we consider a minimum support for the . . . Corporation." Every effort, however, would be made to find this new financing "with an agency and under conditions which will be satisfactory to the Air Force."[29] Shortly thereafter, an approach was made to the Atomic Energy Commission regarding possible areas of AEC sponsorship of RAND research work.[30] It was hoped that the possible AEC support would be on a continuing basis, similar to the arrangement with the Air Force, and that research performed under both contracts could be mutually beneficial to each sponsor.

A subsequent RAND–AEC agreement was worked out along substantially these lines, and it has proved to be a fortunate arrangement for all parties concerned. RAND benefited in being able to keep its research staff intact and in gaining early access to new AEC information relevant to its military research program. The level of support under the AEC contract has been very small compared to the Air Force Project RAND support, but it came at a timely and significant juncture for RAND and through the years has enhanced RAND's research capabilities. The AEC, in turn, has benefited through new knowledge gained under the RAND contract—in particular, on the hydrodynamics of nuclear explosions, on techniques for disarming incoming missile warheads, and on the phenomenology and effects of nuclear detonations. The Air Force has also profited from the RAND-AEC relationship through the enriching of RAND's military research activities

[29] Letter from F. R. Collbohm to Deputy Chief of Staff for Research and Development D. L. Putt.

[30] Letter from F. R. Collbohm to Brig. Gen. James McCormack, Director, Military Applications Division, Atomic Energy Commission, August 18, 1949.

by additional knowledge and insights derived from the AEC work. For example, the AEC contacts played a part in assuring that RAND became aware at an early date of the feasibility of thermonuclear weapons. And as early as the autumn of 1951, before the AEC was prepared to discuss these weapons officially outside the Commission, RAND was analyzing the military implications of this new class of weapons. RAND was able to present the results of its study to the Air Force, the Joint Chiefs of Staff, and President Truman simultaneously with the official AEC announcement regarding the feasibility of thermonuclear weapons.[31]

The AEC contract was not followed immediately by additional sponsorship from other government agencies. The AEC was to remain the only government agency, apart from the Air Force, to share in sponsoring RAND until the late 1950's. At that time, RAND experienced the beginnings of a significant broadening in sponsorship among government agencies that was to include contractual relationships at the Department of Defense level.

An interesting effect of the AEC contract on RAND's internal structure is worth noting. One might expect that the situation created when RAND began to serve two different organizations, with a large area of common interest but also with certain interests at variance,[32] would pose certain adjustment problems for the Corporation. This, in fact, has turned out to be the case but there perhaps have been fewer difficulties than one might have expected. The chief problems have centered on tensions and lack of working contact between the Physics Department and the rest of RAND. Work done under

[31] Internal RAND monograph by Vaughn Bornet, p. 21.
[32] See John W. Finney, "White House Sifts A-Weapons Policy: Tighter Control over Orders by the Military Sought," *New York Times*, September 5, 1962, p. 39, for a discussion of policy differences between the Defense Department and the AEC with respect to the production of atomic weapons.

AEC contract has been performed exclusively by the Physics Department, and as a result that department has become rather set apart from the rest of the RAND research staff. Information has circulated somewhat less freely between the Physics Department and the rest of RAND than between other RAND departments because of the special "Q clearance" required for access to AEC classified materials. Furthermore, the Physics Department occupies a partially sealed-off section of the building to which access is more difficult and more formal than in other parts of the building. All this has led to a somewhat different *esprit* among the Physics Department members and—in the judgment of many RAND people in other departments—to a feeling on the physicists' part that they are a cut above the ordinary RAND mortal. Personal animosities have on occasion developed and, although differences of opinion are common within RAND, some of the deepest and most persistent cleavages have arisen between Physics Department members and other RAND researchers on such questions as the orientation of the research effort, professional standards, and policy recommendations on major issues. It is informal relationships like these that can pose troublesome problems for the research administrator. RAND management always has had to be conscious of artificial barriers that may arise between different skill groups and different organizational units—especially since the organization has operated on a basis of considerable autonomy vested in the individual departments. The tendency for an individual department to withdraw into itself and to build up a self-contained research competence is a particularly vexing management concern.[33]

[33] See Chapter V for a further discussion of problems of internal administration.

RAND AND THE LINCOLN LABORATORY

During the period of RAND's incorporation as an independent nonprofit corporation, there began one other interesting phase in RAND's history that warrants comment. This is the tendency for RAND to have played a direct or indirect part in the formation of additional nonprofit research institutions devoted to defense-analysis work. Early in 1948 the Air Force began to give serious attention to the problem of defending the United States against enemy air attack. The intercontinental bomber was by then a reality and the long-range missile loomed in the not-too-distant future. In May 1948, the Air Force determined that Project RAND would be the best place to carry out an expanded program of intensive research on the air-defense problem.[34] This decision was made largely on the understandable grounds that in Project RAND the Air Force had an already existing research facility that should be utilized to the fullest before any new organization was created. Before long, however, the Air Force came to the conclusion that the kind of research RAND was willing and able to provide on the air-defense question would not meet the requirements and the time schedule that the Air Force had envisaged in originally assigning the task to RAND. It was decided, therefore, to organize a new research facility on a continuing basis to perform this task.[35]

The Air Force then turned to the Massachusetts Institute of Technology for assistance. MIT had on hand the physical facilities (parts of the great wartime Radiation Laboratory complex initially housed the new organization) and the scientific and engineering manpower needed to perform the re-

[34] Letter from Gen. Lauris Norstad to Edward L. Bowles, from the Bowles papers, dated May 18, 1948.
[35] The reconstruction of the sequence of Air Force decisions here is based on materials in the Bowles papers.

search. Thus was born the Lincoln Laboratory with the mission of studying air defense. One can speculate that the Lincoln Laboratory might not have been organized if RAND had more nearly fulfilled the Air Force's expectations on air-defense analysis. More likely, something akin to the Lincoln Laboratory probably would have emerged anyway in light of the growing specialization of function within the research community during the past decade. Unlike RAND, the Lincoln Laboratory's research has been to a considerable extent experimental in nature. RAND continued to do some air-defense research, but of a background and analytical nature. Some of the recommendations issuing from subsequent RAND air-defense studies, however, led to important changes in the armaments used on interceptor aircraft, radar support equipment, data-handling techniques, and other aspects of air-defense operations in the 1953-1957 period.

In the 1950's, RAND was to have a direct role in the establishment of two additional nonprofit research organizations: the Systems Development Corporation (SDC) and ANSER.[36] In addition, a number of RAND staff members with an entrepreneurial bent have resigned and formed "profit" research corporations of their own.[37] One is tempted to ask in

[36] See Chapter IV.
[37] For example, Robert W. Krueger left RAND to found the Planning Research Corporation of Los Angeles; E. H. Plesset founded E. H. Plesset and Associates; Richard C. Raymond set up and for some years managed General Electric's TEMPO Division; and A. M. Mood founded the General Analysis Corporation which was later bought up by the Council for Economic and Industrial Research (CEIR). Other RAND people have left to assume important positions in profit-making analytic concerns which were already in existence, viz., R. M. Salter, Jr. became President of Quantatron, Inc. All of these organizations from time to time hire people away from RAND as one source of recruitment. The most recent RAND "entrepreneur" is Herman Kahn who left RAND in 1961 to found the Hudson Institute in Harmon-on-Hudson, New York. The Hudson Institute, however, is a nonprofit corporation. For a description of the Hudson Institute's operations, see Arthur Herzog, "Report

light of such developments whether there is an internal dynamics or law of bureaucratic proliferation that operates to generate additional research entities.[38] The appearance of such new research organizations in the past decade is probably to some extent the inevitable concomitant of the growing specialization and complexity of the decision-making process. But also to some degree there has probably been a tendency for the particular organizational form of the nonprofit corporation to become "modish."

SUMMARY

By 1950 a mature RAND was beginning to take shape. Initial false starts and growing pains were left behind, and the problems associated with industrial sponsorship were a thing of the past. RAND had built up a competent research staff, and its efforts had reached a "critical mass"[39] point at which research productivity begins to accelerate rapidly. An effective working relationship with the Air Force (and the Atomic Energy Commission) had been established, and a pattern of RAND–Air Force contacts developed that has been more or less followed ever since. RAND would be independent and free to undertake studies of its own choosing, but would honor requests from the Air Force for special research projects and other assistance. RAND would enjoy continuity and stability of support, and would do background research of a broad multi-disciplinary nature directed toward identifying future military requirements and assessing

---

on a Think Factory," *New York Times Sunday Magazine*, November 10, 1963, p. 30 ff.

[38] See Derek J. de Solla Price, *Science Since Babylon* (Yale University Press, New Haven, 1961), for the provocative argument that such expansionist growth is inherent in the nature of scientific research.

[39] The concept of the "critical mass" as applied to research productivity is borrowed from Don K. Price, "Organization of Science Here and Abroad," *Science* 129:759-765 (March 20, 1959).

the implications of developments at the frontiers of science and technology for national security. By 1950, it can also be said, The RAND Corporation began to make its first significant contributions to its major sponsor. As an Air Force General intimately acquainted with RAND affairs said, ". . . it was not until about 1950 before things were coming out of RAND really useful to the Air Force."[40]

[40] Interview with Lt. Gen. Donald L. Putt (USAF, ret.), San Francisco, December 20, 1961.

► IV ◄

A MATURE RAND: 1950–1962

During its early years, RAND functioned almost entirely outside of the public consciousness. The 1950's, however, saw these halcyon days come to an end. The first widespread public notice of RAND's activities came in a March 1951 article published in FORTUNE magazine.[1] The article described in general terms the operations of the RAND Corporation and included a fairly accurate sketch of the project's origins. Since then, a number of additional popular articles have appeared noting the existence of a RAND "think factory" operation as well as a few rather tendentious essays in opinion journals.[2] Most reasonably informed citizens now have some idea of what the RAND Corporation is, and no longer confuse it with the typewriter firm of the same name. A folk song has even been penned about RAND in honor of the folk singer, Pete Seeger, the first and last verses of which run as follows:

> Oh, the RAND Corporation is the boon of the world;
> They think all day long for a fee.

[1] John McDonald, "The War of Wits," *Fortune*, March 1951, pp. 99-102.

[2] See, *inter alia*, "A Valuable Batch of Brains," *Life*, May 11, 1959; Edward Katzenbach, "Ideas: A New Defense Industry," *The Reporter*, March 2, 1961; Joseph Kraft, "RAND: Arsenal for Ideas," *Harper's*, July 1960; David Bergamini, "Government by Computers?" *The Reporter*, August 17, 1961; "Games of Survival," in science section of *Newsweek*, May 18, 1953; Gene Marine, "Think Factory Deluxe," *Nation*, February 15, 1959; "Special Report on Planners for the Pentagon," *Business Week*, July 13, 1963; and Saul Friedman, "The RAND Corporation and Our Policy Makers," *Atlantic Monthly*, August 1963.

They sit and play games about going up in flames;
For counters they use you and me, Honey Bee,
For counters they use you and me.

They will rescue us all from a fate worse than death
With a touch of the pushbutton hand.
We'll be saved at one blow from the designated foe,
But who's going to save us from RAND? Dear Lord,
Who's going to save us from RAND?[3]

This public awareness has brought both advantages and disadvantages for RAND, and has forced the organization to adapt in a number of ways. An initial RAND reaction was the development of what might be called a public-relations awareness in the organization. This awareness was reflected in the evolution of an office responsible for communications and public relations in the early 1950's. The office handles the considerable flow of correspondence that comes to RAND with requests for information about the organization, for copies of RAND studies, and even occasional well-meaning suggestions for matters RAND should study to improve the nation's defense posture. Occasionally, a unique assignment in public relations will arise such as when assorted peace demonstrators gather to picket the building in protest against RAND's complicity in alleged plots against world peace.

[3] Words and music by Malvina Reynolds. © Copyright 1961 by Schroder Music Company. Used by permission.

Characteristically a RANDite penned a rejoinder to Reynolds, the first and last verses of which run as follows:

Oh, the folk-singing crowd is a boon to the world;
They sing all day long all for free.
They sit and sing songs, of injustice and wrongs;
For scapegoats they use you and me, Honey Bee,
For scapegoats they use you and me.

'Twould be nice if our life were only a song,
And as simply resolved as a chord.
That it's not can't be missed; Russian bombs do exist;
It's the fact the folk-singers ignored, Dear Lord,
It's the fact the folk-singers ignored.

Also of interest is the system of deposit libraries that RAND has established in the United States, Canada, and Western Europe beginning in 1953 for the distribution of unclassified RAND materials.[4] This system provides for the dissemination of unclassified RAND materials (except documents for internal use only) to selected library centers for use by scholars and teachers. At present some 2,000 RAND studies are on file in these depository libraries. The Air Force sanctioned the use of Project RAND funds for setting up the initial deposit collections in some 50 libraries in the United States. RAND has financed seven additional deposit libraries in Canada and Europe with its own funds.

RAND's emergence into public consciousness has also resulted in an awakening of congressional interest in the organization. This has led RAND to develop a political sensitivity that to some extent complicates the conduct of its research activities. RAND now has a small office responsible for observing congressional developments, and keeping management and the research staff informed of congressional activities likely to have a bearing on the corporation's operations.

A striking illustration of RAND's suddenly finding itself catapulted into a storm-center of congressional interest is provided by the almost farcical chain of events surrounding the 1958 publication of a RAND study, *Strategic Surrender* (Palo Alto: Stanford University Press), by Paul Kecskemeti. The controversy was touched off when Senator Stuart Symington introduced into the *Congressional Record* on 8 August 1958 an article appearing several days earlier in the *St. Louis Post-Dispatch*. The article alleged that the RAND study (and several others sponsored by the Defense Department) dealt with some future surrender of the United States.[5] An

4 This was done at the suggestion of RAND official Brownlee W. Haydon.
5 See *Congressional Record*, August 8, 1958, p. 15288.

uproar arose from senators protesting against the use of public funds for such a defeatist and unAmerican purpose. When the news reached President Eisenhower, he was reportedly "shocked" and "more excited than at any time since assuming the Presidency."[6] The Department of Defense was forced to conduct a frantic investigation, and issued a statement rebutting the *Post-Dispatch* article. RAND's Washington office issued a press release pointing out the simple facts of the study. (The RAND study was an historical analysis of four instances of surrender in World War II and a critique of the United States' policy of unconditional surrender in these instances. At no point did the study deal with any hypothetical instance of U.S. surrender to a foreign enemy.)

The whole episode came to a ludicrous head with two tense days of heated debate in the Senate on August 14 and 15. It was capped off by the adoption, by a vote of 88 to 2, of a Senate amendment to the Department of Defense Appropriation Bill forbidding the use of public moneys for any study dealing with a U.S. surrender under any circumstances. The nation was treated to the spectacle of senators solemnly protesting that the word "surrender" was not in their vocabulary, and that the very idea of such a study offended against all principles of Americanism and fair play. The President, in the meantime learning the facts of the case, was quoted as saying that "the whole matter is too ridiculous for any further comment."[7] Eventually the facts became known and the furor subsided. The Senate amendment, however, has remained in effect—a grim reminder of the possibility of

[6] "Ike Blows His Top at United States Surrender Article," *New York Mirror* of August 14, 1958, and *Washington Post* article, both reprinted in *Congressional Record*, August 14, 1958, p. 16148.

[7] Senator Richard L. Neuberger, "The Night the Senate Didn't Surrender," *Saturday Review*, September 6, 1958, p. 16.

arbitrary restrictions on the pursuit of objective scholarship.[8]

Beneath the surface frivolity of these events lay a sober warning for RAND. Like it or not, RAND was finding itself propelled into a semi-political role. To some degree it is inevitable that studies, if they are to be "useful" in an immediate sense, will become involved in the political process and be used by various factions contending to have their views crystallized into policy. Nevertheless, a grave danger is presented to an advisory organization like RAND when it becomes widely known and successful and drawn closer to the political arena. This point must be carefully understood. Although such a development is in theory possible, the likely problem is not that the advisory organization will become involved in partisan politics.[9] RAND management has been scrupulously careful throughout RAND's history to avoid identification with partisan causes, and for good reason. The organization's effectiveness almost certainly would have been harmed, and perhaps its existence threatened, if it had been explicitly involved in partisan politics. The real danger is more subtle. The temptation is strong for the advisory organization to become drawn more and more into current policy disputes, thereby shortening its perspective, losing its detachment, and consuming more of its time in "selling" than in

[8] The Russell Amendment (Section 1602, Public Law 85766; 72 Stat. 864; H. R. 13450, "Supplemental Appropriations," approved August 27, 1958) provides: "No part of the funds appropriated in this (or any other) Act shall be used to pay (1) any person, firm or corporation . . . to conduct a study or to plan when and how or in what circumstances the Government of the United States should surrender this country and its people to any foreign power, (2) the salary or compensation of any employee or official of the Government of the United States who proposes or contracts or who has entered into contracts for the making of studies or plans for the surrender by the Government of the United States of this country and its people to any foreign power in any event or under any circumstances."

[9] RAND employees, it will be noted, are not subject to Hatch Act restrictions.

"doing" in the unending quest to achieve a balance between these two.

Indeed, as many a pundit has pointed out, success is a real enemy to the researcher. Once success is achieved, he must spend a considerable portion of his time struggling to avoid the commitments and administrative detail which success often entails. A similar principle may well apply on an organization-wide level. RAND will increasingly face the danger of consuming an unhealthy degree of its research energies in concern with current policy problems, and in playing the role of short-run advisor and handmaiden to a growing list of government officials. The role of RAND is paradoxical: the advisory organization must be known, respected, and listened to in order to make a contribution, but the more it is known, respected, and listened to, the harder for it to be startling, creative, and imaginative. The dilemma of the institutionalized critic is that his inventiveness tends to become increasingly constrained through his desire to secure a hearing for his views.

RAND will also face the simple danger that besets all organizations in the public eye: the necessity for spending a certain portion of its time simply fighting for self-preservation, blunting attacks by enemies, and toning down controversial activities. A striking difference between public and private administration, indeed, is the exceptional degree to which the "public" organization is buffeted by all sorts of pressures and controversies that impose special constraints on the organization's activities and influence the perception of organizational objectives.[10] A danger exists that all other orga-

---

[10] For a discussion of the constraints that beset organizations functioning in the public milieu, see Edward C. Banfield, "Ends and Means in Planning," in Sidney Mailick and Edward H. Van Ness, eds., *Concepts and Issues in Administrative Behavior* (Prentice-Hall, Englewood Cliffs, N.J., 1962), pp. 70-80, and

nizational goals will become subsumed under and subordinated to the goal of self-preservation. RAND will have to be concerned in the future lest the goals it sets merely provide a means for keeping the organization in existence, and intellectual boldness and originality give way to caution and sterility.

All organizations show a concern for self-preservation, but in times of stress this concern may become so pronounced that neurotic reactions result which are actually dysfunctional.[11] In a period of extreme external pressure, for example, it is not difficult to imagine that RAND management might attempt to control the publication by RAND researchers of unclassified articles in the open literature or seek to regulate closely the research staff's freedom to give speeches to academic, industrial and other non-RAND audiences. Realism suggests that there is some legitimate management interest in these areas. But it is precisely this sort of management control that is potentially most dangerous for an organization like RAND. Unwise application of even informal constraints could invite great indignation among key members of the research staff and possibly the resignations of important individuals. Or the organization could easily turn into a "tame" arm of its clientele.[12] There is some informal feeling within RAND that the organization has grown less "free" in recent

Harry Eckstein, *The English Health Service* (Harvard University Press, Cambridge, Mass., 1958), ch. ix.

[11] For a discussion of the consequences of external conflict for social groups, see Lewis A. Coser, *The Functions of Social Conflict* (Free Press, Glencoe, 1956), esp. pp. 87-110. Coser's analysis, however, is intended generally as an examination of the positive role that conflict may play in social relations. Whether conflict will have adverse or beneficial consequences for an organization or other social group will apparently depend on a variety of factors, including the intensity of conflict and the internal structure and cohesion of the group.

[12] A classic treatment of an organization becoming to some extent a "captive" of its clientele through the mechanisms of cooptation is Philip Selznick, *TVA and the Grass Roots* (University of California Press, Berkeley, 1949).

years, and that management has gone about as far as it can legitimately go in overseeing the research staff's external contacts. One minor manifestation of neurotic behavior already in evidence is management's overconcern with the RAND "image." A RAND foray into the realm of institutional advertising is an unhappy recent case in point. In the period from 1955 to 1960, RAND spent $225,132.11 on institutional advertising in *Scientific American, Foreign Affairs*, trade journals, and a number of college and university journals. This served no purpose other than image-building, for RAND's recruitment practices are not geared to institutional advertising.[13] The practice, however, has since been discontinued. In the end, the only "image" that matters for an organization like RAND is its reputation for quality research.

Finally, RAND's move into public awareness and the awakening of congressional interest in the organization have opened up some interesting possibilities in regard to legislative-executive relationships. The possibility of "end runs" around an executive sponsor to Congress could make for intricate and troublesome problems in the sponsor-advisor relationship. RAND is acutely, perhaps overly, conscious of the dangers of this tactic; it has strived to avoid any congressional contacts either at the staff or management level that would lessen sponsor confidence in the organization. RAND does, however, maintain a number of informal contacts with congressmen and the staffs of various congressional committees. On occasion, RAND staff members have done studies at the request or suggestion of congressmen or committee staff.[14]

[13] For critical comments regarding this practice, see James McCamy, *Science and Public Administration* (University of Alabama Press, Birmingham, 1960), p. 58.

[14] Study Paper No. 18, "National Security and the American Economy in the 1960's," by H. Rowen, January 30, 1960, prepared for the Joint Economic

The most ambitious effort RAND has undertaken for Congress to date is the compilation of a space handbook for the House Committee on Space and Astronautics. This effort, done as a public service with RAND's own research funds, earned the plaudits of a number of congressmen and doubtless served a useful function from RAND's point of view in winning some new friends in Congress. The handbook was subsequently commercially published in several languages.

CONGRESSIONAL ATTITUDES TOWARD THE NONPROFIT ADVISORY CORPORATION

Congress as yet does not seem to have a fully crystallized outlook toward the nonprofit advisory corporations. Congressional attitudes here can perhaps best be described as divided and uncertain. In 1961 the Subcommittee on Department of Defense Appropriations of the House Committee on Appropriations appeared to put itself on record as strongly critical of the nonprofit corporation.[15] Its position tended to lump together all the different nonprofit organizations into one category with little awareness that, say, RAND and Aerospace might be markedly different entities. The House Committee on Government Operations, in contrast, has seemed more favorably disposed toward the nonprofit corporation concept. Indeed, it was instrumental in the Air Force's creation of

---

Committee, 86th Cong.; Senate Document No. 101, "The Nature and Feasibility of War and Deterrence," by H. Kahn, included as Exhibit 1, in *Organizing for National Security*, Hearings before the Subcommittee on National Policy Machinery of the Committee on Government Operations, U.S. Senate, 86th Cong., February 23-25, 1960; "Report on a Study of Nonmilitary Defense," Exhibit B in *Civil Defense*, Hearings before the Subcommittee on Military Operations of the Committee on Government Operations, House of Representatives, 85th Cong., May 1958; and R. Buchheim, et. al., *Space Handbook: Astronautics and Its Applications*, Staff Report of the Select Committee of Astronautics and Space Exploration, 85th Cong., December 29, 1958, prepared at the request of Hon. John W. McCormack.

15 See Chapter I, note 22.

Aerospace, the nonprofit corporation that functions as a management aid for the Air Force System Command.[16] The Committee on Government Operations also appears to have an understanding of the different uses to which the nonprofit corporation can be put and an awareness of the different kinds of nonprofit corporations that the government has created.

It appears likely that Congress will manifest a growing interest in the advisory corporations.[17] It is interesting to speculate on the possible implications for RAND of increased congressional interest. RAND's concern for congressional relations will probably intensify, and a new and more extensive pattern of RAND–Congress contacts perhaps may emerge. RAND's traditional advisory position within the executive branch has always posed the difficult problem of an "end run" around a sponsor to a higher policy echelon. But the prospect of congressional involvement in the internal policy debates set off by a defense agency's confidential advisor would surely present issues of a far greater complexity. It seems possible, in any event, that the future will bring an intricate interplay of advisor-executive-legislative relations

---

[16] The Committee on Government Operations, in *Organization and Management of Missile Programs*, Eleventh Report of the Committee on Government Operations (House Report 1121), September 2, 1959, p. 99, had this to say about nonprofit corporations: "The subcommittee believes that if STL [Space Technology Laboratories] is to have any future with the Air Force, it must be converted into a nonprofit institution akin to the RAND Corporation and other private and university-sponsored organizations which serve the military departments and other agencies of the Federal Government on a stable and continuing basis. Government relationships with nonprofit organizations also pose problems, but they are less important than the benefits received, and certainly less crucial than those posed by the STL tie with the Air Force."

See also *Air Force Ballistic Missile Management (Formation of Aerospace Corporation)*, Second Report of the Committee on Government Operations, 87th Cong., 1st Sess., House Report 324, May 1, 1961, and *Organization and Management of Missile Programs*, Hearings before a subcommittee of the Committee on Government Operations, House of Representatives, 86th Cong., 1st Sess., February 4, 5, 6, March 2, 3, 5, 12 and 20, 1959.

that will pose some subtle issues and provide a fascinating glimpse of the governmental process.

RAND DEVELOPS A STRATEGIC SENSE

Accompanying RAND's move from relative obscurity into the public limelight was another broad development of considerable importance in the 1950's. This was RAND's development of a "strategic sense." In the early years, as previously seen, RAND studies tended to be engineering efforts or else analyses of rather low-level problems akin to what operations researchers did in World War II. The studies were rather elaborately mathematical in nature, and showed little concern for integrating a number of complex variables, some qualitative in nature, into a broad context of some future "system" whose contours and implications in terms of military effectiveness can only be dimly foreseen. Studies of "military worth" in RAND's early years also showed little awareness of strategic implications, even though they were supposed to be concerned with the broader implications of military techniques and instrumentalities. For example, a typical early study employing the concept of military worth dealt with the effects on bomber safety and bombing accuracy of releasing bombs at higher and lower altitudes.[18]

17 Congressional inquiries that have touched directly or indirectly on advisory corporation operations include: *Systems Development and Management*, Hearings before a Subcommittee of the Committee on Government Operations, House of Representatives, 87th Cong., 2nd Sess., parts 1-5, 1962 (Holifield Subcommittee Hearings) ; *Tax-exempt Foundations and Charitable Trusts: Their Impact on Our Economy*, Committee Print, Chairman's Report to the Select Committee on Small Business, House of Representatives, 87th Cong., 2nd Sess., December 31, 1962; *The Aerospace Corporation: A Study of Fiscal and Management Policy and Control*, Report of the Subcommittee for Special Investigations of the Committee on Armed Services, U.S. House of Representatives, 89th Cong., 1st Sess., August 12, 1965.

18 J. D. Williams, RA-15008, "The Effect on Military Worth of Exchanging Bombing Accuracy for Bomber Safety by Increasing Range of Bomb," September 1, 1946.

During the 1950's, however, RAND sharply departed from a preoccupation with engineering studies and low-level operations research analyses. In the first place, a number of the nation's outstanding strategic thinkers have developed at RAND. Such men as Bernard Brodie, Herman Kahn, Albert Wohlstetter, William W. Kaufmann, and Thomas C. Schelling have all made important contributions to strategic thought while either RAND staff members or consultants.[19] Strategic thought, as an activity in its own right, has become an important part of the RAND research program.

Secondly, an awareness of strategic implications has penetrated into other parts of the RAND research program, particularly the broad-gauged systems-analysis work that has become a vital part of the RAND research program. Something of a revolution took place in the 1950's which transformed the typical RAND systems analysis from a narrowly technical product into a novel application of numerous professional skills to a broad policy problem.[20] The Strategic Base Study done in the early 1950's provides a good illustration of a systems-analysis project in which strategic considerations begin to play a large role in the analysis.[21] Concerned

[19] See, for example, Bernard Brodie, *Strategy in the Missile Age* (Princeton University Press, Princeton, 1959; Herman Kahn, *On Thermonuclear War* (Princeton University Press, Princeton, 1960); Thomas C. Schelling, *The Strategy of Conflict* (Harvard University Press, Cambridge, Mass., 1961); William W. Kaufmann, ed., *Military Policy and National Security* (Princeton University Press, 1956); and Albert Wohlstetter, "The Delicate Balance of Terror," *Foreign Affairs*, January 1959, and "NATO and the Nth plus 1 Country Problem," *Foreign Affairs*, April 1961. This work, it should be recalled, represents only unclassified contributions.

[20] For an elaboration of this point, see R. D. Specht, "RAND—A Personal View of Its History," *Journal of the Operations Research Society of America* 8:836-837 (November-December 1960), and Herman Kahn, *On Thermonculear War*, pp. 119-126.

[21] See Chapter VI below. The first important RAND systems analysis probably was a 1948 Offensive Bomber Study by a team headed by E. Paxson. The study recommended the adoption of a turbo-prop airplane as a future

with showing the most efficient basing system for SAC bombers for a period five years hence, the Base Study became involved in such important strategic issues as the vulnerability of overseas bases to enemy nuclear attack. This called into question certain fundamental assumptions of existing Air Force strategic doctrine.

The point, however, needs to be stated carefully. Not everybody at RAND is concerned with strategy or policy in a broad sense. Much of RAND's work remains research and technical advisory services of a more "traditional" nature. Indeed, work within the traditional disciplines continues to provide the raw materials on which the useful systems analysis must build. The number of true "generalists" at RAND— those concerned with integrating the fruits of specialized research into a broad policy context—has probably never exceeded fifteen percent of the professional staff at any given time.[22]

This strategic consciousness serves to distinguish RAND from some of the other advisory corporations whose major concern is more or less exclusively with nonstrategic matters.

---

strategic bomber, a recommendation the Air Force did not accept. Objecting to certain design features of the proposed airplane, a ranking Air Force general is said to have dismissed the study with the comment: "I wouldn't strap my ass to that RAND bomber under any circumstances." The next significant systems analysis was a 1951 Air Defense Study by E. L. Barlow and colleagues. This study had moved closer to the "real world," and had some impact on the Air Force's air defense posture in the 1953-1957 period. See Chapter III, p. 90.

[22] David Novick, head of RAND's Cost Analysis Department, once estimated the figure as 25 out of 464 professionals on the RAND staff. Quoted in Wesley W. Posvar, "The Impact of Strategy Expertise on the National Security Policy of the United States," *Public Policy*, XIII (Yearbook of the Graduate School of Public Administration of Harvard University; 1964), p. 48n26. Posvar estimates that of the combined staffs of six major advisory corporations, numbering something less than two thousand professionals, perhaps only a hundred "can be regarded as specialists in strategy at the national policy level; the remainder are scientists and technologists who work on elements of our defense posture without primary regard for the whole." *Ibid.*, p. 48.

The research activities of the Human Relations Research Office (HumRRO) and the Special Operation Research Office (SORO), for instance, deal almost entirely with tactical matters (akin to the operations-research activities of World War II), with training techniques and devices, and with man-machine relationships.[23] ANSER, also, does only little in the way of research on broad strategic questions, being mainly concerned with providing short-run staff assistance to the Directorate of Development Planning, Headquarters USAF. It does, however, deal with aspects of future Air Force plans and technical-feasibility studies of future weapons and as such can be said to have a partly strategic character. In any case, its research activities are broader in nature than HumRRO's or SORO's. The training concern of the latter organizations make them almost semi-developmental agencies.

The Research Analysis Corporation (RAC), on the other hand, does research with strategic implications, including studies on "finite" deterrence and the concept of limited war, along with other narrower and more specialized aspects of its research program. Examples of the latter include studies on such subjects as tactical combat situations, weapons for infantrymen, troop mobility, detailed supply problems under various assumptions of warfare, and communications on the battlefield. But RAC does much less research of a broad strategic character than RAND, partly for the reason that RAC's sponsor, the Army, in the past has been typically less directly concerned with the strategic aspects of national security than the Air Force.

---

[23] Based on interviews with Dr. Lynn Baker, Army Behavioral Research Office, September 12, 1961; Dr. Charles Hutchinson, Human Relations Research, U.S. Air Force, September 14, 1961; Dr. Charles Bray, the Smithsonian Institution, September 15, 1961 (the interviews were all held in Washington, D.C.);

The Operations Evaluation Group (OEG; since 1962 a part of the Center for Naval Analysis) had until the late 1950's been more or less oriented toward the traditional operations-research problems and techniques of World War II. Since then the Navy has made an effort to augment its advisory capabilities, and has sought to develop a capacity for long-range strategic planning and analysis along with the traditional operations-research effort.

The Institute for Defense Analyses (IDA), of the remaining nonprofit advisory institutions, seems most nearly to resemble RAND in being concerned with broad strategic issues as a vital part of the research program. This is perhaps hardly surprising since IDA is attached to the Department of Defense level in the defense establishment—the level which is, in theory, concerned solely with "policy" and not with "operations."

Despite the apparent strategic interest one finds in certain other defense-advisory organizations, one is struck by the general "flavor" of RAND work. In contrast to the sister organizations, RAND studies seem often to be more truly provocative, to present evaluations of genuine strategic alternatives, and to present views that cut across particular agency or service jurisdictions. The work of the other advisory organizations, even when it presumably involves broad strategic considerations, generally seems less inventive and more constrained by prior commitment to prevailing policy assumptions. The independent RAND "atmosphere" in general seems more conducive for the emergence of truly creative ideas in policy research than the working atmosphere of the advisory unit closely tied to the sponsoring agency.[24]

---

and on *HumRRO Research Bulletin* 8, "What HumRRO Is Doing," August 1961. George Washington University, Human Resources Research Office.

[24] See Chapter VII.

SOME PROMINENT RESEARCH PAYOFFS IN THE 1950's

Some idea of the scope and expansion of RAND research activities in the 1950's may be gained from a brief mention of some prominent research "payoffs" during this period. A first area, just discussed, is the field of strategic thought. Second, RAND studies have made contributions to scholarship in the area of Soviet studies. RAND has been a pioneer in this field and its work has been of benefit not only to RAND sponsors, but to the scholarly community at large.[25] A third

[25] The unclassified part of RAND's work in this area includes such titles as: Nathan Leites, *The Operational Code of the Politburo* (McGraw-Hill, New York, 1951); Nathan Leites, *A Study of Bolshevism* (Free Press, Glencoe, 1953); Nathan Leites and Elsa Bernaut, *The Ritual of Liquidation: The Case of the Moscow Trials* (Free Press, Glencoe, 1954); Merle Fainsod, *Smolensk Under Soviet Rule* (Harvard University Press, Cambridge, Mass., 1958); Margaret Mead, *Soviet Attitudes Toward Authority: An Interdisciplinary Approach to Problems of Soviet Character* (McGraw-Hill, 1951); Raymond L. Garthoff, *Soviet Military Doctrine* (Free Press, Glencoe, 1953); Herbert S. Dinerstein and Leon Gouré, *Two Studies in Soviet Controls: Communism and the Russian Peasant and Moscow in Crisis* (Free Press, Glencoe, 1955); Herbert S. Dinerstein, *War and the Soviet Union: Nuclear Weapons and the Revolution in Soviet Military and Political Thinking* (Praeger, New York, 1959); Myron Rush, *The Rise of Khrushchev* (Public Affairs Press, Washington, D.C., 1958); V. D. Sokolovskii, *Soviet Military Strategy*, translated and annotated by Herbert S. Dinerstein, Leon Gouré, and Thomas W. Wolfe (Prentice-Hall, Englewood Cliffs, N.J., 1963); Thomas W. Wolfe, *Soviet Strategy at the Crossroads* (Harvard University Press, Cambridge, Mass., 1964); Myron Rush, *Political Succession in the USSR* (Columbia University Press, New York, 1965); Philip Selznick, *The Organizational Weapon: A Study of Bolshevik Strategy and Tactics* (McGraw-Hill, New York, 1952); Leon Gouré, *The Siege of Leningrad* (Stanford University Press, Stanford, Calif., 1962); Arnold Horelick and Myron Rush, *Strategic Power and Soviet Foreign Policy* (University of Chicago Press, Chicago, 1966); and Roman Kolkowicz, *The Soviet Military and the Communist Party: Institutions in Conflict*, forthcoming 1966. RAND studies of Soviet science policy include: Arnold Kramish, *Atomic Energy in the Soviet Union* (Stanford University Press, Stanford, Calif., 1959) and F. J. Krieger, *Behind the Sputniks: A Survey of Soviet Space Science* (Public Affairs Press, Washington, D.C., 1958). RAND contributions to an understanding of Soviet economics include: Abram Bergson and Hans Heymann, Jr., *Soviet National Income and Product, 1940-1948* (Columbia University Press, New York, 1954); Walter Galensen, *Labor Productivity in Soviet and American Industry* (Columbia University Press, New York, 1955); Oleg

area where RAND research has made some contributions to knowledge and had some tangible "payoffs" as well is in the realm of missile and satellite technology. New knowledge of weapon design advances, together with years of research on guidance, re-entry heating, and power plants by RAND, assisted the Air Force significantly in its decision to proceed with the large-scale development and production program for the first Atlas ICBM. Further, many aspects of the Discoverer, Samos and Tiros development programs have been based on RAND research. In recognition of some of this work, the American Meteorological Society in January 1961 gave a special award to two RAND scientists "for their pioneering work in the planning of a meterological satellite."[26] A RAND study, also, was credited by the Air Force with contributing to a substantial reduction in the initial program cost of the Titan ICBM.[27] In addition, Project RAND made a formal recommendation with detailed design suggestions for a defense missile system to counter the threat of a low-altitude enemy attack. The Air Force turned this recommendation over to the Army, which then undertook the present HAWK program. These are a few examples of important RAND contributions in the field of missile and satellite technology. For further examples of unclassified RAND studies in this area, the reader is referred to the voluminous *Index of Unclassified*

Hoeffding, *Soviet National Income and Product in 1928* (Columbia University Press, New York, 1954) ; Abram Bergson, *The Real National Income of Soviet Russia Since 1928* (Harvard University Press, Cambridge, Mass., 1961) ; Janet G. Chapman, *Real Wages in Soviet Russia Since 1928* (Harvard University Press, Cambridge, Mass., 1963) ; and Richard Moorsteen, *Prices and Production of Machinery in the Soviet Union, 1928-1958* (Harvard University Press, Cambridge, Mass., 1962).

[26] Internal RAND document, "Some Major Examples of RAND Research," May 1961. The recipients of the award were W. W. Kellog and Stanley Greenfield. The work for which they received the award was done in the early 1950's.

[27] Internal RAND materials, dated 1960.

RAND Publications, located in deposit libraries across the country. It should be remembered, however, that the unclassified publications represent only a part of RAND's total effort here. A significant body of classified literature also exists which is available only to cleared personnel in defense agencies and defense-contract firms.

Fourth, several new mathematical tools, one invented and largely developed at RAND and the other to whose development RAND has contributed—dynamic and linear programming—have constituted an important RAND contribution to scholarship and to the solution of specific military problems.[28] Applications of these techniques have been made in aircraft time-to-climb calculations (minimizing time or fuel consumption), communications-network-flow studies (maximizing capacity of communications lines), logistics problems (inventory and allocation decisions, production smoothing, cargo loading, and the like), satellite-trajectory problems, equipment-replacement policy, the analysis of the atmospheres of other planets, the equilibrium composition of complex chemical mixtures, and the assignment of repair personnel and equipment at missile site and depot to maximize equipment readiness, routing and re-routing of communication networks under attack, and automatic-weapons assignment. Further, dynamic and linear programming have had wide use in American and foreign industry. The U.S. petroleum refining industry, for example, universally employs a RAND linear-programming code in its product-mix calculations.

Other esoteric analytical tools have also been largely developed, or their use perfected, at RAND. The concept of

[28] Richard Bellman, prolific RAND mathematician, played a large role in the development of dynamic programming. George Dantzig, the father of linear programming, spent some years at RAND in the 1950's perfecting the theory he had developed earlier. He eventually left RAND to join the staff of the University of California at Berkeley.

system analysis itself—the application of diverse techniques and skills to a broad problem—is largely a RAND product (though somewhat related aids to management decision had been used for some years within large industrial enterprises).[29] RAND has made important contributions to the development of such esoteric and precise mathematical techniques as Monte Carlo and game theory as well. These highly abstract intellectual activities occasionally can have practical uses. The H-bomb, for example, probably could not have been designed without the aid of Monte Carlo techniques. And the formal models of game theory sometimes can clarify salient aspects of "real world" behavior, and "sensitize" the researcher to various kinds of reactions that one might expect from an individual or nation in certain decision contexts. This in turn can serve as a very useful stimulus to other parts of the RAND research program.

An activity somewhat related to formal game theory—war gaming and simulation—has also been useful in varying degrees in helping to clarify certain kinds of military problems. More important, perhaps, is simulation's use in serving as a heuristic device for stimulating further research on various problems revealed as significant in the course of the game play or simulated situation.[30] Gaming and simulation

[29] An excellent exposition of the systems-analysis approach and its intellectual antecedents may be found in Roland N. McKean, *Efficiency in Government Through Systems Analysis: With Emphasis on Water Resource Development* (John Wiley, New York, 1958), esp. pp. 5-21. See also Charles J. Hitch and Roland N. McKean, *The Economics of Defense in the Nuclear Age* (Harvard University Press, Cambridge, Mass., 1960), and Herbert A. Simon, *The New Science of Management Decision* (Harper, New York, 1960).

[30] A lively, and sometimes acrimonious, controversy has sprung up within RAND over the merits of gaming as a research tool. For a general description of RAND's political gaming, see Hans Speier and Herbert Goldhammer, "Some Observations on Political Gaming," *World Politics* 12:71-83 (October 1959). A balanced treatment which fairly sums up the strengths and weaknesses of

also have had important uses as a training device for government officials to help them understand what kinds of behavior to be prepared for in various crisis situations. Crisis games became widely used by high State and Defense Department officials early in the Kennedy Administration, and the reaction of participating officials was almost unanimously favorable. In sum, even if RAND had made no substantive contributions to the solution of specific problems, it would still be fair to say that RAND's pioneering work on new tools of analysis has been a valuable "payoff."

Another area that deserves mention is RAND's contribution to the theory and practice of program budgeting.[31] Program budgeting promises to become a management tool of great importance in an era of big government and large enterprise. Already the "McNamara revolution" in Pentagon management practices has made extensive use of program-budgeting concepts developed earlier at RAND, and a number of former RAND staff members have played leading roles in implementing the McNamara reforms.[32] It is likely that program-budgeting techniques will eventually be used by other federal agencies, although there may be some formidable difficulties in implementing budgetary reforms in certain mission areas.

gaming is Sidney Verba, "Simulation, Reality, and Theory in International Relations," *World Politics* 16 (April 1964).

[31] See these works by David Novick: *Efficiency and Economy in Government Through New Budgeting and Accounting Procedures* (The RAND Corporation, R-254, February 1954) ; *A New Approach to the Military Budget* (The RAND Corporation, RM-1795, 1956) ; *Program Budgeting in the Department of Defense* (The RAND Corporation, RM-4210, September 1964); *Program Budgeting: Program Analysis and the Federal Budget* (Harvard University Press, Cambridge, Mass., 1966).

[32] For a general description of the "McNamara revolution," see William W. Kaufmann, *The McNamara Strategy* (Harper, New York, 1964) and Theodore H. White, "Revolution in the Pentagon," *Look*, April 23, 1963, pp. 31-49. Notable among the former RAND staff members who have played leading roles in the McNamara reforms are Charles J. Hitch, Henry E. Rowen, and Alain Enthoven.

An important step toward government-wide use of program budgeting occurred on August 25, 1965 when President Johnson announced that he had directed Budget Bureau chief Charles L. Schultze to proceed with plans for a new planning and budgeting system for all federal agencies beginning with the budget for the fiscal year 1968.[33] Although the President did not specifically mention the Defense Department, it was clear that he intended the budgetary reforms initiated under Secretary McNamara as a model for other executive agencies.

A last area I will mention as an important RAND research payoff relates to studies of the new structural materials titanium and beryllium. Titanium (and its alloys) was found to be a useful structural material in research conducted at the Battelle Memorial Institute under RAND subcontract. Although the work was done in the first few years of RAND's existence, it did not begin to find its major applications until the 1950's. The feasibility of using beryllium as a structural material was demonstrated at RAND in the mid-1950's to a large extent through the work of RAND engineer, Dr. George A. Hoffman. The conception and development of Hoffman's research project led to his forecasting a potentially spectacular improvement in airplane and missile performance through the use of beryllium as a major structural material.[34] This

[33] "Introduction of New Government-Wide Planning and Budgeting System," Statement by the President to Members of the Cabinent and Heads of Agencies, August 25, 1965. For brief commentaries, see "Johnson Altering U.S. Cost Control: Says Revolutionary System Will Be Used for Budget," *New York Times*, August 26, 1965, p. 17, and "Budget Wins New Role," *Business Week*, August 21, 1965, pp. 128-132. Henry E. Rowen, who left the Defense Department in 1964 to join the Bureau of the Budget, and William Capron, another former RAND economist, have been important figures in the Budget Bureau's efforts to stimulate government-wide interest in program budgeting.

[34] George A. Hoffman and William R. Micks, *A Re-evaluation of Beryllium as a Potential Structural Material for Use in Flight Vehicles*, The RAND Corporation, RM-1642 (May 7, 1956), and Hoffman, "Beryllium as an Aircraft Structural Material," The RAND Corporation, P-871 (this article was published

RAND research helped make it possible for designers of aircraft, missile and space vehicles to contemplate using the metal beryllium with its better weight-strength ratio at high temperatures in place of metals like aluminum, thus greatly improving performance and ultimately reducing costs of vehicle development.

The above are but a few examples of important payoffs resulting from the Air Force's investment in Project RAND. The contributions cover a number of different areas, and vary from the largely individual efforts to the general analyses in in which large groups of researchers participate. The value of these several types of contributions is hard to quantify. But there can be little doubt that handsome dividends have resulted to the Air Force and to society at large for the original investment of public funds in Project RAND research.

THE SPIN-OFF OF THE SYSTEM DEVELOPMENT CORPORATION

During the 1950's, the RAND staff grew noticeably. At one point in 1957, the total number of RAND employees reached a maximum of 2,605.[35] Part of this expansion in staff came about as a result of the addition, at Air Force request, of a significant logistics-research capability to the RAND effort, particularly in the supply, data-processing, and transportation fields. By September 1955, the RAND Logistics Project was well under way, and the subsequent provision of additional funds by the Air Force led to the establishment of

---

in part or in its entirety in such journals as the *Aeronautical Engineering Review*, February 1957; *Aviation Week*, December 17, 1956; *Materials and Methods*, February 1957; *Western Aviation*, December 1956; and *Western Metals*, December 1956).

[35] The figure of 2,605, however, is rather misleading. A large part of the figure can be attributed to the growth of a systems training activity which, as we shall see, was gradually disassociated from RAND and eventually became incorporated as a separate nonprofit organization, the System Development Corporation.

what became known as the Logistics Systems Laboratory. This activity has remained a part of the RAND research program. However, a more important cause of the expansion of RAND's staff was the mushrooming growth of a systems-training activity that threatened at one point to change the whole nature of the RAND research program. RAND's recognition of the dangers of becoming too deeply involved in an activity contrary to its long-range research mission is the story of the creation of SDC (System Development Corporation).

In the early 1950's, some RAND psychologists began to study how groups of men operating complex machines work under conditions of stress. They chose an Air Defense Direction Center for detailed study. As part of the research design, a simulated center was created to test the reactions of individuals when called upon to perform increasingly complex tasks under various conditions of simulated enemy air attack. At this early stage, the activity was thought of solely as a research project. As the project progressed, however, it became evident that the research findings would have important implications for the training of Air Force personnel operating Air Defense Direction Centers. Typical junior college students were brought in as an experimental group, and subjected to various tests requiring them to respond to blips on radar screens as though a hostile raid were in progress. As an aid to learning, each amateur crew was given a special opportunity to rectify its mistakes and to plan improvement immediately after conclusion of each test series. The results of the research showed that the student crews learned much faster and were able to cope with much heavier enemy raids than the most experienced Air Force crews. The implications of similar training techniques for Air Force personnel were obvious. When General Frederick H. Smith, Vice-Commander of the Air Defense Command, saw the research results he quickly concluded that RAND had discovered a new training method which should be put to prompt use. Thereupon the activity broadened into a separate System Development Division at RAND which was assigned a number of new tasks and considerably augmented by additional Air Force funds.

By the mid-1950's, RAND management began to realize that they were facing a situation of the tail threatening to wag the dog. The System Development Division had grown almost twice as large as the rest of RAND, and further growth seemed inevitable as the Air Force contemplated the assignment of more new tasks to the division. Moreover, the division had departed from its original function of research on man-machine relationships. By this time it had become almost entirely a large training activity performing a routine technical service for the Air Force and responding to short-run requests on a crash basis. RAND management felt that continuance of this activity under RAND direction might distort the organization's whole character, and decided that a way must be found to disassociate the activity from RAND. The RAND Board of Trustees and RAND management also felt that the contribution the new activity could make in the national interest would be greater if it were permitted to grow and develop on its own.[36]

Initially, some consideration was given to a plan whereby the System Development Division would be spun off as an industrial corporation to engage in research on man-machine relationships for the Air Force, to provide training assistance, and perhaps eventually to produce actual training devices for

[36] As the then SDC President M. O. Kappler explained to the Holifield Subcommittee (Hearings, part 3, p. 991): "They [the RAND trustees] felt that the realization of the potential inherent in this new area required the formation of an independent organization to concentrate in it. This area, as well as RAND's primary mission, would both be served best by the creation of an independent nonprofit corporation to carry on the work of the division."

sale on the market.[37] The plan was dropped, however, when the Air Force strenuously objected. Fearful of conflict-of-interest problems, General Smith stated the Air Force's unwillingness to accept the creation of an industrial corporation out of the System Development Division. He indicated that the only alternative arrangement the Air Force would consider would be another not-for-profit corporation patterned along RAND lines. Accordingly, RAND management began to make plans to form a separate nonprofit corporation out of the System Development Division. Most of the division had gradually been moved to nearby rented facilities, and the activity was being managed apart from the normal RAND research operations. The necessary arrangements were now undertaken to give the division a separate legal status and its own managerial structure. By the end of 1957 these arrangements were largely completed and on 1 December 1957 the division was spun off as a separate nonprofit organization under the name System Development Corporation. The new organization became established in separate quarters two miles away from the RAND building in Santa Monica.

Initially, RAND provided substantial administrative support and guidance to the corporation. The officers and trustees of the corporation were partly RAND people, and RAND officials assisted the SDC in drawing up its contract with the Air Force. But gradually RAND withdrew its support and the SDC became fully independent. In January 1961, at the request of the Bureau of Federal Credit Unions, RAND severed even its credit union ties with the SDC, the last remaining administrative connection with the new corporation.

[37] Confidential interviews. California law provides that a nonprofit corporation may engage in profit-making ventures so long as all income derivable from the venture is expended for the purposes stated in the corporate charter and does not inure to the benefit of any individual. Federal law, however, is less permissive on this point.

In 1962, the only formal remaining link between RAND and the SDC was severed when RAND Vice President J. R. Goldstein left the SDC Board of Trustees. There remain, however, a number of informal contacts at the staff level between RAND and SDC as between RAND and other research groups.

Since its spin off from RAND, the SDC has grown to nearly four times the size of RAND in staff, and its budget is nearly three times the size of RAND's.[38] SDC's work has been mainly in the area of training and planning for the most efficient use of complex man-machine systems; an important major function has been the SDC role in helping the Air Force establish and operate the SAGE system. Its concern with the interaction of man and machine and with training devices has lent a distinctive "hardware" flavor to the SDC research program that contrasts sharply with RAND's research program.

RAND deserves credit for recognizing the dangers that the systems-training activity presented to its research mission. Such instances are rare of an organization spurning an opportunity to acquire the trappings of additional power and wealth in the interests of adhering to a central goal. Perhaps a healthy measure of prudence also contributed to RAND management's decision to divest the organization of the systems-training activity. The large research organization is normally the one that is most conspicuous and most likely to arouse hostile criticism, as well as presenting a more likely target for budget cuts by economy-minded congressmen and budget officers. Moreover, the rapid growth that would have

[38] SDC later was to have an indirect hand in the establishment of another Air Force nonprofit corporation—the MITRE Corporation, a systems engineering organization working for the Air Force's Electronic Systems Command. MITRE performs a similar kind of function for the Electronics Systems Command as Aerospace does for the Air Force's Systems Command—technical supervision and program guidance. There is some interlocking of membership on the SDC and the MITRE Board of Trustees.

been associated with the systems-training activity was incompatible with the RAND policy of maintaining close working relationships among a relatively small group of high-caliber researchers. An overly rapid growth might have jeopardized the established pattern of interactions among RAND researchers. In the years since the SDC spin-off, RAND has continued to grow but at a level gradual enough to permit the new staff generally to become acclimated to the RAND "subculture."[39] Thus the new researcher tends to adapt to the mores of the organization rather than the character of the organization drastically changing under a sudden influx of new people. However, as we shall see, RAND's growth has brought with it certain problems, notably in regard to flexibility and internal communication.

THE FOUNDING OF ANSER

The SDC was not the only nonprofit organization in whose founding RAND participated in the 1950's. Not long after the spin-off of the SDC, RAND became involved in the creation of a second group—Analytic Services, Inc. (ANSER) —and played a leading role in nurturing this infant organization until it was able to stand on its own feet. The story of ANSER's establishment is interesting from several points of view. Before the story can be understood, however, it is necessary to go back to the early 1950's and pick up the chain of events from there.[40]

[39] The growth rate was approximately ten percent per annum until the early 1960's. A leveling off then occurred. RAND now seems to have more or less stabilized around the level of 1100 employees with additions to the staff approximately equal to the departures. The normal "turnover" or departures from RAND's professional staff has been about five to seven percent per annum.

[40] The following account of ANSER's development is based on interviews with Dr. Stanley Lawwill, President of ANSER; L. J. Henderson, Jr., RAND Vice President and Chairman of the Executive Committee of ANSER's Board; F. R. Collbohm, RAND President; and Crawford Thompson, assistant to the RAND Treasurer. Any interpretations drawn from the events are my own.

In 1951, an Assistant for Evaluation office was established within the Air Staff (the name was later changed to Director of Development Planning). The office's primary responsibilities were to assist in estimating the technical feasibility of new weapons, planning the Air Force's research and development objectives, and (as an Air Staff arm) to be generally responsive to immediate needs of the Air Staff. From the start the office was understaffed. Great difficulty was encountered in getting either an adequate number of military officers with appropriate technical training or an allotment of PL 313 positions to bring in civilian technical skills from the outside. As one means of helping the office fulfill its responsibilities, RAND was asked to provide some technical people "on loan" to the office. RAND complied with the request, but the arrangement did not work out very satisfactorily and was later discontinued.

The Director of Development Planning was also provided with certain funds to contract for studies with outside research organizations. A number of contracts were let to such groups as the Cornell Aeronautical Laboratory and the Stanford Research Institute. A Washington, D.C., consulting firm, Corvey Engineering, was hired on contract initially to print and edit for distribution the various studies produced by the other contractors. Gradually, Corvey Engineering assumed the role of monitor and prime contractor for the various technical studies administered by the Directorate of Development Planning. The Air Force subsequently decided that it would be preferable to pool resources and build a single study group instead of parceling out small fragments to various research institutions. The Corvey Engineering group was a natural candidate for the job, having already developed an effective working relationship with Air Force personnel.

In the meantime, however, Corvey Engineering had been bought up by the Melpar, Inc., a manufacturing firm engaged in the business of developing and selling test equipment and other items of hardware to the Defense Department. Melpar is actually a subsidiary of the Westinghouse Air Brake Company. Corvey Engineering had thus become a division of a manufacturing company. Notwithstanding, a contract was consummated with the Melpar Company in September 1957 for the creation of a Scientific Analysis Office formed around the nucleus of old Corvey Engineering personnel. Not long afterward, other Air Force industrial contractors and the military people in the Air Research and Development Command (now the Air Force Systems Command) objected to the arrangement on grounds that it presented a possible conflict-of-interest situation. It became apparent that the Melpar contract would have to be terminated and some alternative arrangement made.

An interesting parallel can be drawn between the Melpar experience and RAND's experience when it was still attached to the Douglas Company. Like Douglas, Melpar sought to wall off the Scientific Analysis Office from the rest of the firm and to administer the project separately from the normal company activities. And also as in the Douglas case, there was no evidence that Melpar benefited or attempted to benefit unfairly from its position as confidential advisor to the Air Force. Yet the suspicions remained; and again the concept of a continuing government-advisory unit being attached to a firm in the hardware business proved unworkable. Further, like RAND when it was housed at Douglas, the Scientific Analysis Office at Melpar encountered certain irritants and difficulties as a result of being associated with an engineering firm. Procedures designed for blue-collar people and development work again proved inappropriate for a research organization. This factor also contributed to the desire to find a new home for the Scientific Analysis Office.

Early in 1958 the Director of Development Planning

launched an intensive search for some organization either to take over the Melpar Scientific Analysis Office and administer the project on a continuing basis or to build up anew a similar capability. After a nationwide survey of research organizations, the Directorate concluded that the RAND Corporation would be the organization most qualified to handle such a project. RAND reentered the picture at this point when invited in the summer of 1958 to take over the Melpar unit.

RAND management considered that such a project was not consistent with the regular RAND method of operation and so informed the Air Force. RAND objected that the task envisaged was primarily of a short-range nature, and would require a responsiveness to routine Air Staff needs. RAND was not organized or staffed for such a function. In informal conversations, RAND people suggested that perhaps what the Air Force needed was to strengthen its "in-house" technical capability since what they envisaged seemed to be really a kind of auxiliary staff function.

The Air Force, however, declined to accept the latter suggestion. The most likely in-house analytic facility that might have been able to perform the assignment—the civil service Operations Analysis Office—had unfortunately fallen into disrepute in many parts of the Air Force. (The Operations Analysis Office is composed of some 200 civilian technical personnel trained largely in mathematical techniques of operations research; many of these individuals work directly with Air Force commands in the field.) The Operations Analysis Office had suffered a deterioration in the quality of its professional staff, and it was also widely believed to have become too closely preoccupied with Strategic Air Command matters to provide objective technical advice to other parts of the Air Force. Undoubtedly, these factors contributed to the reluctance of the Development Planning people to use in-house analytic facilities to replace the Melpar unit. RAND, in consequence, was urged to reconsider its decision and to establish a RAND division to take over the Melpar function. RAND management decided that RAND could best assist by helping to establish a new nonprofit organization akin to (or actually built around) the Melpar Scientific Analysis Office. The Air Force Development Planning office after some hesitation agreed to this course of action. Shortly thereafter, in July 1958, ANSER was founded under RAND sponsorship as a new nonprofit research organization incorporated under the laws of California.

In general, RAND exercised close initial supervision over the new corporation. Originally, all of ANSER's corporate officers were RAND people except Dr. Stanley Lawwill, the man who had recently been selected to head the Melpar group and who was chosen to head the new corporation. Lawwill was well known to RAND management, having served with distinction in Air Force operations research since World War II. RAND lent the new corporation working capital, and entered into a management-services agreement in which RAND agreed to keep all administrative records for the new corporation until ANSER was equipped to perform the task itself. Gradually, RAND lessened its administrative control and phased out the management-services agreement. Finally, on 12 January 1961, ANSER became independent from RAND direction; the only remaining formal link between the two groups since then has been the presence of two RAND executives on the ANSER Board of Trustees. ANSER remains the "baby" of the nonprofit research corporations, having only a staff of some 40 professionals. Its offices are located in Virginia not far from the Pentagon.

ANSER's mission, as indicated earlier, is of a considerably different nature from that of RAND. It involves analytical cost-effectiveness studies and technical evaluations of weapons systems and subsystems, and the provision of technical advice to the Directorate of Development Planning on various day-to-day problems. In practice, the latter function has come to be of primary importance. The bulk of the ANSER effort has tended to be channeled into short-range projects and fulfilling requests for immediate staff assistance from the Directorate for Development Planning. Also, ANSER's studies, unlike RAND's, are not to a significant degree self-generated. To a limited degree, ANSER can be said to initiate studies by means of making suggestions for possible areas of fruitful inquiry to the Directorate of Development Planning. These suggestions may then come back to ANSER in the form of directives to undertake given studies. But the difference between RAND and ANSER in this regard is graphically illustrated by the requirement for Air Staff approval before ANSER can adjust its research priorities or proceed on a study.[41]

Several interesting issues are posed by ANSER's founding and subsequent operations. One has to do with the ANSER policy of tailoring its salary opportunities closely to existing civil service salary schedules. Because of the location near Washington and conscious of an awakening congressional interest in the nonprofit corporations, ANSER management made a firm decision from the start not to invite possible criticism by offering salaries to its professional staff substantially more attractive than civil service opportunities. This practice contrasts rather sharply with the policies of RAND and some other institutions in the nonprofit research community. And it raises the interesting question as to whether ANSER's policy will form a precedent for the other nonprofit

[41] Information on Air Staff approval of research priorities obtained from interview, Dr. Stanley Lawwill, February 1, 1962, Fairfax County, Virginia.

research corporations. Could this practice, if widely adopted, help point the way toward a *modus operandi* between the nonprofit advisors and the civil service? Or could ANSER afford to adopt the policy only because it was not bidding for the most sought after talent on the market?

Second, it is clear that ANSER was created because there were no adequate "in-house" analytic facilities to perform the task which ANSER now performs for the Air Force's Directorate of Development Planning. The Air Force as early as 1951 had relied on contractual assistance from outside institutions for this task. The question arises: in principle would it have been wiser to have built up the analytic facility within the government itself? The question appears especially pertinent since the task was essentially a short-range auxiliary staff type function. We shall return to these complex questions at a later point.

RAND AND DIVERSIFICATION IN SPONSORSHIP

As previously noted, RAND acted historically as confidential advisor to one client—the Air Force—although in 1950 it broadened its sponsorship to the point of doing a modest level of research for the Atomic Energy Commission. Beginning in 1956, however, a significant change has occurred in the traditional pattern of sponsor support. RAND has markedly broadened its base of sponsorship until it has now reached the stage of performing a wide variety of research tasks for a number of different sponsors.

In 1956, RAND began the first of two small task orders it has undertaken at the request of another government agency; the projects involved research on ruble-dollar ratios for representative types of construction. The level of effort involved in these projects was small; and the dollar-values of the two contracts amounted to only $25,104 and $56,736.

125

Then in 1959 the really significant steps toward diversification of sponsorship began after the Air Force announced its intention of freezing the Project RAND support at its then current dollar level. As in the case of the original effort to diversify in 1950 with the AEC contract, financial considerations rather than doctrine played the major role, initially at least, in RAND management's decision to seek additional sponsorship.

A contract consummated with the Advanced Research Projects Agency in 1959 became the first significant effort toward diversifying both in terms of the dollar amount of the contract and the nature of the new sponsorship. The ARPA contract initially was to run for a period of one and a half years and provide support amounting to a total of $1,500,000 (later the time period was extended and new funds were added). With this contract, outside sponsorship obviously had begun to reach a point where it could be important to RAND's over-all financial outlook.

The nature of the research work done under the ARPA contract did not differ greatly from what RAND was already doing under Project RAND and the AEC contract. The ARPA work was to be "theoretical, conceptual and background studies and analyses of advanced military weapons systems and components." Still, the ARPA contract occasioned some unrest in the Air Force in contrast to RAND's contractual relationship with the AEC (the Air Force had in general endorsed and approved of the AEC contract). Now, for the first time, RAND had assumed the role of researcher and advisor for an agency at the Office of the Secretary of Defense (OSD) level. RAND was thus, in effect, working for both the Air Force and the Air Force's superior in the Defense hierarchy—the latter having the power of life-and-death over the service through the assignment of roles and missions to the component parts of the defense establishment. To many Air Force people this seemed to compromise RAND's position as confidential advisor to the Air Force.

Also, the new contract raised fears that the magnitude of the research effort envisaged might lead RAND to divert an unduly large amount of research time away from specifically Air Force problems, leaving the Air Force in the position of continuing to pay a major share of the bills while receiving less attention to its own needs.

These suspicions were heightened when, several years later, RAND moved into a whole new series of contractual relationships at the OSD level. Not long after the Kennedy Administration took office, a contract was entered into with the Office of the Comptroller within the Office of the Secretary of Defense to provide RAND assistance in the establishment of a new defense budgetary system. The new system, drawing heavily on concepts developed at RAND and put into effect to a large extent through the efforts of Mr. Charles J. Hitch, new Defense Comptroller and former senior RAND economist, was to revise drastically traditional budgetary procedures in the Defense Department. Thereafter, in early 1962, a large contract was consummated with the Office of the Assistant Secretary of Defense for International Security Affairs (ISA). The ISA contract involved analytical studies of a variety of defense problems, including counter-insurgency and limited war questions, and the annual funding under the ISA contract for a two-year period amounted to over $1,000,000. The ISA contract frightened the Air Force more than any other because many Air Force officers felt that some of the civilians in ISA were contemptuous of military professionalism and sought to downgrade the professional soldier's status.

After the projected freeze in funding announced in 1959, RAND also with the Air Force's blessing entered into contracts with the National Aeronautics and Space Administration (NASA). The contracts were of intermediate dollar amounts, and ranged from broad-gauged research on the long-range social and economic implications of space activities to very specific studies of a carefully circumscribed problem formulated by the agency.

It should also be noted that RAND assumed several new contractual relationships with the Air Force separate and apart from Project RAND in the period of the late 1950's and early 1960's. Contracts were negotiated with the Air Force Technical Applications Center, the Air Force Office of Scientific Research, and Headquarters, Ballistic Missiles Center of the Air Materiel Command. The work to be done included linguistic research and costing studies of launching vehicle systems.

Finally, in 1961 RAND also added the National Science Foundation and the National Institutes of Health to its growing list of clients, obtaining a one-year contract from the former for a study of meteorological conditions and prospects of weather modification and from the latter a small contract for research on mathematical aspects of certain biological phenomena. And additional research contracts were soon to be consummated with such new sponsors as the Agency for International Development (AID) and the Defense Atomic Support Agency (DASA).

RAND had also received grants from the Ford Foundation, the Social Science Research Council, and the Carnegie Corporation for studies, respectively, of urban transportation, heuristic programming, and simulation of cognitive processes.

Thus in the period of several years RAND was transformed from an organization working principally for one major sponsor under a single broad contract into an organization with many clients and a multiplicity of diverse contracts. After the diversification moves RAND had some 22 separate contracts and four grants with almost a dozen different clients.[42] Project RAND support, while still the major part of RAND's budget, dropped from 95 percent to 68 percent of the corporation's total support as a result of the diversification moves.

Two major consequences have followed from taking on the additional contractual obligations. One is the growing complexity of RAND's internal administration. There is some feeling within RAND that the organization has become more "bureaucratic" than it has ever been. The second major consequence is the strain that diversification his imposed on RAND's relationship with its major sponsor; some important changes in the nature of that relationship can be traced in part to the diversification moves. The two consequences are interrelated and mutually reinforcing.

First, RAND had been spared many troublesome administrative problems in the past since it had only one sponsor to please, continuity of support, and a broadly-worded and flexible contract to work under. The AEC contract altered the situation somewhat, but basically did not pose many new administrative difficulties for RAND. The management problems began to emerge more significantly when RAND was faced with the task of keeping track of research in progress for numerous different sponsors, scheduling work in such a way as to insure that all contract obligations were being met, and in general engaging in more planning of the research

---

[42] RAND has also done some research on sub-contract for such organizations as the Livermore Radiation Laboratory and the Jet Propulsion Laboratories of the California Institute of Technology. Since these projects are under a master contract for the Atomic Energy Commission and for the National Aeronautics and Space Administration, they are not counted as separate clients in this tabulation.

effort. The job of drawing up and negotiating a new contract also involves administrative time and energy, and a certain amount of promotional activity is often necessary to interest a potential client in a contract in the first instance.

In addition, RAND had in the meantime grown in size which has helped to make the very informal management policies of an earlier era less applicable to its current situation. Some research on organizational behavior suggests that increasing size, combined with age and functional complexity, may lead to a growth in the staff or overhead function relative to the line or operating function of an organization.[43] The present findings offer some support for this hypothesis. RAND has shown a tendency to be more conscious of management problems and to devote more staff time to administration relative to research as the organization has grown larger and more complicated. See Figure 1.

The fact that RAND has been functioning in a more hostile external environment may also help to explain the increased "bureaucratization" of the organization. It may be

[43] Frederic W. Terrien and Donald L. Mills, "The Effects of Changing Size upon the Internal Structure of Organizations," *American Sociological Review* 20:11-13 (1955); Theodore Anderson and Seymour Warkov, "Organizational Size and Functional Complexity," *American Sociological Review* 26:23-28 (1961); Peter M. Blau and W. Richard Scott, *Formal Organizations* (Chandler, San Francisco, 1962), pp. 123-127; and on the significance of age as contributing to an accretion of rules and regulations in social structures, see Wilbert E. Moore, *Social Change* (Prentice-Hall, Englewood Cliffs, N.J., 1963), pp. 25-26, 58. There is a lively controversy in the literature on the relative significance of these various factors as determinants of increases in administrative activity within organizations. In the present case, it is difficult to make a discriminating judgment. RAND became larger and more complex (that is, assumed numerous new contract obligations for a variety of clients) at roughly the same time— and in the meantime was growing older. In part, also, the increases in RAND's administrative staff may be accounted for in terms of the "formalization" process which all organizations seem to undergo sooner or later. For an elaboration of the conceptual difficulties in analyzing these factors, see William H. Starbuck, "Organizational Growth and Development," in James G. March, ed., *Handbook of Organizations* (Rand McNally, Chicago, 1965), pp. 451-533.

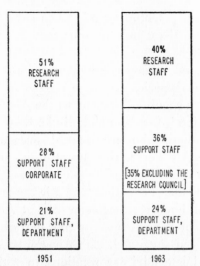

Figure 1. A comparison of administrative and research staff in 1951 and 1963; from data compiled by R. D. Specht, RAND Research Council

supposed that an organization in some fashion "senses" its external environment. In periods of external pressure, there seems to be some tendency toward increased centralization of authority in organizations and greater emphasis put on explicit rules and procedures.[44] The interaction that one may hypothesize between an organizational unit and its external environment, however, is a rather difficult notion that is in need of conceptual clarification. Similarly, one must be cautious in speaking of the presence or absence of "bureaucratic" elements unless that concept is given a clear operational definition. What can be said for present purposes is that diversification in sponsorship, reinforced by such factors as increasing size and the emergence of congressional critics

[44] Bernard Berelson and Gary A. Steiner, *Human Behavior: An Inventory of Scientific Findings* (Harcourt, Brace, New York, 1964), p. 370, and Charles F. Hermann, "Crisis and Organizational Viability," *Administrative Science Quarterly* 8:61-82 (June 1963). Cf., however, Lewis A. Coser, *The Functions of Social Conflict*, pp. 87-110.

and other real or potential enemies, has led to more formal management practices and to somewhat less flexibility in internal operation.

The second and related consequence of diversification is that RAND has experienced some degree of estrangement from its primary sponsor.[45] Of course, the Air Force like any large organization is far from monolithic. RAND has friends, enemies, and those who are neutral toward it scattered throughout the Air Force. But the attitudes of disaffection described here are deep and pervasive enough to warrant using the short-hand abstraction "the Air Force" to represent the opinions of important segments of the Air Force hierarchy. Several of the reasons for this partial estrangement have already been suggested. The Air Force, seeing RAND develop an intimate advisory relationship with agencies at the level of the Office of the Secretary of Defense, has had second thoughts about its confidential lawyer-client tie with RAND. At the least, some Air Force officers now seem to believe that they must be considerably more circumspect in their dealings with the corporation. There is, further, a concern about communication of RAND studies done for the Air Force to higher echelons of the government. The Air Force can find itself having to answer to higher authority about why it has not adopted a RAND recommendation by virtue of RAND's extended avenues of communication throughout the higher reaches of the federal government. This can be a serious irritant. It has doubtless added to the grievances that many in the Air Force have felt toward particular RAND staff members who have been zealous in "selling" the results of their research to high-level policy

---

[45] The discussion in this section is based on conversations with Air Force and Department of Defense officials and individuals in the nonprofit research community.

makers. Also, the Air Force has understandable reservations about the prospect of continuing to provide the major part of RAND's funding while an ostensibly ever-increasing share of the RAND research effort is being diverted away from specifically Air Force concerns. In general, it can be said that there is a widespread feeling in the Air Force that RAND has "grown away" from its original sponsor.

These difficulties and critical sentiments have been magnified by the fact that the Air Force no longer occupies the undisputed dominant position in the American military establishment. In many ways, with a somewhat stable strategic balance already achieved or in prospect, the Air Force has faced a decline in its relative importance. The emphasis on such concepts as limited war and counterinsurgency have tended to some extent to thrust the Army back into the limelight.

Moreover, cutting across the question of relative service primacy is the whole shift of power and influence in the defense establishment away from the individual services to the unified commands and to the agencies at the Department of Defense level headed by an increasingly powerful Secretary of Defense.

The impact of these developments on RAND–Air Force relations is evidenced in such matters as greater Air Force sensitivity to RAND studies critical of current Air Force doctrine, a desire for closer monitoring of RAND publications policy, impediments to the free flow of information to RAND, and some disappointment that RAND has not been more eager to understand and support the Air Force's position on key issues that seem to involve the Air Force's whole future. RAND has, of course, always had disagreements with the Air Force in the past on Air Force polices and procedures. A principal RAND function, indeed, is to act as an informed and friendly critic of current Air Force practices from a long-run perspective. But, as the Air Force's position has become more vulnerable, policy disagreements with RAND have at times seemed to take on alarming proportions with some Air Force people. RAND's failure to support the Air Force position in the B-70 controversy, for example, was a particular sore point with some Air Force elements.

It was the cumulative effect of various RAND–Air Force differences that led to RAND's budget for the fiscal year 1961 being cut in half to $7,000,000 in the initial round of budget preparation.[46] The cut was eventually restored before the Department of Defense budget was presented to the Congress. Had the cut come into effect, a drastic and perhaps fatal reduction in staff and the scope of RAND activities would have been forced upon RAND management. On other occasions during the 1957-1962 period, the Air Force's Strategic Air Command (SAC) has in effect temporarily broken off diplomatic relations with RAND. Still another reflection of strains and stresses in the RAND–Air Force relationship appeared in the Air Force Policy Statement on Non-Profits issued in late 1961 over the signature of the Secretary of the Air Force, Eugene M. Zuckert.[47]

The statement indicated, among other things, that the Air Force would not consider depreciation costs to be properly chargeable to Air Force contracts in the case of contractors owning their own facilities. This declaration sent a shiver of apprehension through RAND officials. For RAND had decided early to enter on an extensive construction program rather than pay high rentals (a reimbursable cost under the

---

[46] Confidential interviews with officials in the Office of the Comptroller (Office of the Secretary of Defense) and the U.S. Air Force.

[47] The statement is reprinted in *Air Force and Space Digest*, December 1961, p. 79.

contract) for office space or seek government building funds through facilities contracts. Depending upon how rigidly the Air Force contracting officers interpret the policy statement, RAND's operations could be seriously affected as considerable amounts of money are involved in the depreciation allowance. In sum, it can be seen that RAND's relations with the Air Force have entered into a new phase. The relatively tranquil days of the past are now largely gone, and a more difficult relationship has emerged. RAND and the Air Force find themselves in something of the position of an older married couple with the honeymoon over—yet without sufficient cause to consider a separation. Both remain conscious of the mutual interests and mutual benefits tying them together, but each partner has now grown somewhat chary of the other.

On the issue of diversification, and the strains in the Air Force relationship of which it is a partial cause, at least two distinct schools of thought have emerged at RAND. One holds that diversification should be pursued as a conscious policy so that RAND would not be overly dependent on any single sponsor and could operate from a more independent base *vis-à-vis* all clients. The Bell Report gave the supporters of this view a boost by endorsing the principle of diversification in sponsorship for the not-for-profit corporations engaged in policy research.[48] The other school of thought maintains that the long-standing close advisory relationship with the Air Force, clearly of enormous benefit to the corporation in a variety of ways despite its occasional stresses, should not lightly be compromised. Due regard is paid under this view

[48] The report stated: "Organizations of this kind [operations and policy research organizations like RAND] should not be discouraged from dealing with a variety of clients, both in and out of Government." Holifield Subcommittee Hearings, part 1, p. 227.

to the fact that the Air Force continues to provide about 70 percent of RAND's total support, despite the recent sharp increase in non-Air Force sponsorship. The latter view has apparently prevailed for the time being with top RAND management. In testimony before the Holifield Subcommittee in 1962 RAND President Franklin R. Collbohm came out strongly in favor of continuing to rely on primary Air Force sponsorship:

Mr. Roback (staff investigator): Do you consider, Mr. Collbohm, that RAND will continue to be in the foreseeable future, as it has been in the past, primarily an Air Force contractor?
Mr. Collbohm: We hope so.
Mr. Roback: You are not interested in building up a diversified operation as such?
Mr. Collbohm: No. As a matter of fact, I do not think it would work too well. I think the Air Force has—well, I was going to say had developed over the years—but I retract that and say it started right out with a philosophy and a policy as to how to handle the type of an organization that RAND is, that is practically perfect, I would say. And it would be very, very undesirable for the country as a whole, if this relationship should be changed.[49]

There are clearly some strong arguments in support of RAND management's position. Diversification in some ways seems to pose more uncertainties than it offers reasonable assurance of strengthening the research program and safeguarding the corporation's future. Continuity of support and enlightened sponsorship have certainly been instrumental in RAND's success as a research organization. Indeed, stability of sponsorship is a significant advantage that an organization like RAND enjoys over, for example, an organization like the Stanford Research Institute (SRI) which has many clients and a large number of small contracts. RAND is spared much of the promotional activity and "brochuremanship" that

49 Holifield Subcommittee Hearings, part 3, p. 952.

besets the SRI in the latter's perennial search for new business from a host of clients. Moreover, SRI faces a number of problems of research administration that RAND does not, at least to the same degree, in regard to scheduling work, assigning personnel to specific projects, and meeting deadlines on a large number of small individual contracts. An organization like the SRI may also be somewhat disadvantaged with respect to the hiring of professional people. Since hiring tends to be tied to the success or failure of efforts to secure new contracts, there is a certain inflexibility in recruitment practices. RAND can more easily avoid sudden increases or decreases in staff, and can afford to be more flexible in recruiting high-caliber staff.

Further, RAND's experience thus far with its new civilian clients has apparently not indicated that more enlightened sponsorship is necessarily to be found with civilian agencies. On the contrary, there is some feeling in RAND that most of RAND's new civilian clients have had less experience in research administration than the Air Force and consequently have sometimes demonstrated what from RAND's point of view seems to be an imperfect understanding of the mechanics of research work. Task orders have been unduly specific, deadlines have been too rigid, and there has been on occasion some tendency to expect more of research than it can reasonably offer. All of which has added force to the cautionary arguments within RAND against moving rapidly away from primary Air Force sponsorship.

Considered from the point of view of RAND's place within the Defense Department hierarchy, there are several additional objections that can be raised to further diversification. To take on more and more research and advisory tasks at levels higher than the Air Force would draw RAND inevitably closer to the center of defense policy making, and probably

137

would contribute to fundamental changes in the organization's character. Realism suggests that the advisory organization closely tied to the top policy echelons would encounter difficulties in gaining access to sensitive information and in general would be more subject to client pressures. One might see a shortened time perspective for studies, less concern with long-range research with no immediate application, more of a tendency to blow with the current political and doctrinal winds, and a greater emphasis on the short-run service function at the expense of the research function. To some extent, changes such as these have already taken place in RAND. It is a keenly felt danger among many RAND people that the organization has begun to exhaust its intellectual capital, and at this point needs to devote more attention to acquiring new knowledge and less attention to applying what is already known. In this connection it is instructive to reflect on the experience of the Institute for Defense Analyses (IDA). It has been harder to make a success of IDA in part because its close ties with the top defense-policy echelons have resulted in the organization having less freedom to define the frame of reference of studies. RAND's position close to, yet somewhat aloof from, the centers of power seems to be a more promising setting for a successful research operation.

Although there are strong arguments against a further broadening of RAND sponsorship, it is clear that continued reliance on primary Air Force support has some drawbacks as well. One difficulty that is growing steadily more acute is the problem of continuing to attract a first-rate staff at a time when Air Force concerns no longer are the central focus of the nation's defense policies. As previously noted, RAND's recruitment had benefited in the past from the broad Air Force mission which was almost equivalent to the whole of the nation's defense effort. RAND has thus begun to find itself in a curious position: it has sought to convince the Air Force of the continued importance of the traditional advisory relationship at a time when good people are increasingly difficult to recruit for work on Air Force problems. The same trend is evident with established RAND staff members. Many are less interested in working on Air Force problems because Air Force problems in some ways seem less exciting than they were in the past. The net result has been a possible slight decline in the quality of RAND's staff in certain skill areas since the end of the 1950's.[50] The most difficult aspects of RAND's staffing problem are probably two: attracting senior people from the universities and keeping the able younger people who tend to leave after relatively short periods in the organization.

It is apparent that diversification would offer certain advantages to the organization in providing a greater variety of research opportunities for the professional staff. One can thus probably anticipate some strong pressures in the future for additional diversification, either at the Office of the Secretary of Defense level or from some other source (other government clients or possibly a foundation). Similar pressures may well operate with respect to the other service-connected advisory institutions. A root cause will be the steady accretion of power and influence over the past decade at the broad Department of Defense level, and away from the individual services. The organization that does not adapt to this trend in some way seems to face the danger of drifting into a state of decline.

[50] As a senior RAND researcher remarked in an interview: "Before, when I went up to Department X, I had a feeling I was talking to the A students; now when I go up there, I get the feeling I'm talking to B+ students." This sort of question, however, is difficult to judge in any systematic way. In studying human organization, one may often encounter a "good-old-days" syndrome in which the past is glorified and present problems are magnified.

CONCLUSION

In the early 1960's RAND had reached a curious stage in its development. In one sense, the organization had never been more successful. RAND had moved from a small, largely unknown group into the public consciousness in a meteoric fashion that startled old-time RAND people. RAND generally enjoyed a high prestige in the research community. In some quarters, the name of RAND evoked something akin to veneration. RAND alumni were scattered widely throughout industry, government, and the academic community, leaving RAND with a number of friends in high places and a voice in important policy decisions difficult to imagine a decade earlier. New clients sought out RAND's services in certain subject areas where the corporation had a proven competence. Many RAND personnel served on advisory committees, boards, and consultant groups of the federal government. In the years 1959 and 1960 alone, there were more than 200 instances of RAND personnel serving in such capacities.[51]

At the same time, however, certain disquieting signs were beginning to appear. RAND's very success and prominence brought with it serious disadvantages for the conduct of the research effort and helped make the organization a target for hostile criticism. RAND had been forced to develop a political sensitivity, and to be more cautious in its dealings with clients and the outside world in general.

Congress had taken an increased interest in the nonprofit corporations, and it appeared likely that a closer congres-

[51] Fifteenth Annual Report, March 1, 1961, pp. 32-48. These assignments included such varied tasks as membership on the Department of Defense Scientific Advisory Committee for Ballistic Defense, participation in the International Federation of Automatic Control in Moscow as U.S. delegate, special consultant services to the Bureau of the Census, and attendance at the Conference of Advisors to the Pugwash Meetings.

sional surveillance of organizations like RAND would result. President Kennedy's high-level study of the whole spectrum of government contracting for research and development (the Bell Report) became available publicly on 1 May 1962.[52] Although it was the fundamental conclusion of the report that "it is in the national interest for the Government to continue to rely heavily on contracts with non-Federal institutions to accomplish scientific and technical work needed for public purposes," the report noted that "where the contracting system does not provide built-in controls . . . attention should be paid to the reasonableness of contractors' salaries and related benefits."[53] It went on to suggest, among other things, possible steps toward exercising a tighter control over the salaries paid to persons working under a research and development contract from a government agency.[54] The report also devoted considerable attention to the conflict-of-interest question, and strongly suggested that more effective, although as yet undefined, measures were needed to safeguard the public interest against conflict-of-interest situations for institutions as well as for individual consultants and scientific advisors.[55]

In addition to the increased government interest, there has been a growing concern within the industrial community about the supposed inroads the nonprofit corporation is making into industry's sphere of interest.

[52] *Report to the President on Government Contracting for Research and Development* [The Bell Report], reprinted in the Holifield Subcommittee Hearings, part 1, pp. 191-337. For useful commentaries on the Bell Report, see J. Stefan Dupré, "The Efficiency of Military Research and Development: Kaysen, Cherington and the Budget Bureau," *Public Policy*, XII (Yearbook of the Graduate School of Public Administration of Harvard University; 1963), pp. 286-301, and "Administration By Contract: An Examination of Governmental Contracting-Out," *George Washington Law Review* 31:685-783 (April 1963).

[53] Bell Report, in Holifield Subcommittee Hearings, part 1, p. 239.

[54] *Ibid.*, pp. 239-240.

[55] *Ibid.*, pp. 223-227.

In brief, the external environment in which RAND operates has changed considerably over the past decade. Powerful political forces have arisen to question the assumptions under which RAND and other groups were founded. A generally more questioning and skeptical attitude seems to have arisen. All of which increases the likelihood that the years ahead may be more anxious ones for RAND, and suggests the possibility that a larger fraction of RAND energies may have to be spent in fence-mending activities—an uncomfortable posture for a research organization.

Moreover, RAND has in a sense worked itself out of a job to some extent over the past decade. RAND, in serving an important function as a kind of training center for people to acquire new analytical tools for the study of complex problems, has spread knowledge and understanding of its research techniques to other institutions and to other parts of society. Many professional people have served with RAND for several years and then departed, bringing to their new position the capabilities and experience acquired at RAND. RAND has also gone to considerable lengths to help the Air Force and other clients develop capabilities in such areas as cost analysis, weapons evaluation, and advanced budgetary techniques. This has served to downgrade RAND's importance in a relative sense as the capability of other groups has increased, and as what were once considered recondite analytical tools have become more widely known and practiced.

In this context RAND's relationship with the universities is an interesting case in point. In the immediate postwar years, there were no significant coordinated programs of research on national security issues underway in any of the major universities. RAND was in a near monopoly position *vis-à-vis* the universities with respect to defense-studies research for some years. What was done was largely RAND-inspired or in terms of a frame of reference established at RAND. A corollary of this was that top minds interested in research in the field of national security could be more easily attracted to RAND because of the concentration of talent already existing there. Subsequently, however, a number of universities have begun to establish seminars and study programs dealing with defense matters. The Harvard Defense Policy Seminar was the pioneer program of this kind but was followed quickly by similar programs at such universities as Princeton, Columbia, Chicago, Ohio State, Wisconsin, UCLA, MIT, Johns Hopkins, and Dartmouth College.[56] Currently, these universities are producing much valuable research on national-security issues. The awakening of academic interest in national-security problems has resulted in diminishing to some extent RAND's prominence as a major center for national-security research. On occasion in recent years important RAND researchers have left to join the staff of university departments. The growing difficulties of attracting qualified staff may force RAND to rely more and more on part-time consultant arrangements with persons in the academic community. RAND's character would certainly be changed if this were to happen on any substantial scale. RAND would become less of a research organization and more of a holding company to funnel talent toward work on a range of public problems.

Furthermore, from the point of view of RAND's internal operations, we see some acute problems beginning to emerge which will increasingly tax the imagination and ingenuity of RAND officials. For one thing, the corporation has grown to a point where internal communication is no longer easy. It is increasingly difficult to know what is happening in the vari-

[56] Joseph Kraft, "The War Thinkers," *Esquire*, 58:102-104 (September 1962).

ous different parts of the organization. Interdisciplinary research at RAND, far from becoming steadily easier and more effective, has actually grown more difficult in recent years. The individual departments in some respects seem to have grown more self-contained, inward-looking, and conscious of their own prerogatives, and less interested in working on interdisciplinary projects. To the above, one must add the complicating factor of diversification which has multiplied the problems of managing RAND's extensive research enterprise. As RAND has found itself in a more hostile external environment, this too has had an effect on the organization's internal management.[57]

In the face of increased pressures for closer administrative controls over the corporation's activities, perhaps the greatest challenge confronting RAND management is the task of preserving under changed conditions what has been RAND's most valuable asset—the freedom of the researcher to do

[57] One possible effect, as discussed above, is the prospect of closer supervision and tighter administrative control by top management in periods of stress for an organization. In terms of its impact on the research program, this would mean that a study would tend to undergo more painstaking internal review before being released to the client or the public. In an effort to adduce quantitative proof for this statement, I hypothesized that the percentage of "D's"—or internal working papers—may have increased relative to other categories of RAND documents. A comparison of "D's" and "RM's" at three different time intervals, however, was inconclusive. The average monthly output of "D's" did increase 280 percent from 1950 to 1963 as compared to a 167 percent increase of "RM's" over the same period. But most of that percentage increase in "D's" occurred from 1950 to 1955—the period *before* the upsurge of critical scrutiny of the advisory corporations. Nevertheless, it seems to be true that in recent years RAND publications have been subjected to more rigorous pre-publication criticism and review before being released to the client or the public. The principal categories of RAND documents are the following: R's—Report (classified or unclassified monographs); RM's—Research Memoranda (classified or unclassified shorter monographs. Both the R's and RM's are publications for clients); P's—Papers (journal-type unclassified professional contributions); and D(L)'s—Limited Documents (same as D's, internal technical communications and working papers, except that the distribution even within RAND is controlled by the author; classified or unclassified).

objective studies and analyses with a minimum of interference from arbitrary deadlines, administrative or operating responsibilities, or concern with preconceived policy lines. To some at RAND the conscious quest to preserve the conditions allowing for freedom seems a dangerous interference with the mystique of the organization. Actually, such a view seems no longer applicable to the RAND situation. The increasing self-consciousness and self-appraisal that one notices at RAND is probably necessary if the organization is to find answers to the complex issues it will likely face throughout the next decade. The organization has simply grown too large to permit reliance on the same informal methods of intercommunication and management that prevailed in an earlier day. Further, diversification has imposed new demands upon management: it must assure that the work product is responsive to the specific needs of particular clients. The pitfalls and sometimes onerous chores that success entails require that special care must be taken to insulate the productive researcher from excessive demands on his time and energy. But if RAND administration is not to defeat is own ends in the process, it must be extraordinarily imaginative in coming years. It must develop to near perfection the art of the research administrator acting as "pump" rather than "bottleneck."[58]

The freedom that RAND enjoyed in its early history, moreover, may be undergoing a subtle yet far-reaching transformation. As an organization, RAND may perhaps be less "free" and "independent" in the sense that it may have to be more responsive to specific client requests and less able to rely on a broad open-ended research contract from one or

[58] Sir Eric Ashby, "The Administrator: Bottleneck or Pump?" *Daedalus*, Spring 1962, issue devoted to Science and Technology in Contemporary Society, pp. 264-278.

two major sponsors. Yet, from the perspective of the individual researcher, freedom can continue to exist insofar as responsibility for a study or project will be assigned internally only to those interested in working in the particular area. For those who need occasional relief from work on any of the current client-oriented projects, the opportunity remains open to carry on RAND-sponsored research of one's own choosing.[59] Thus RAND management acts as a kind of buffer, cushioning the impact of client requests on the working staff by parceling out given research assignments to those it can interest and persuade to accept them, and in general shielding the researcher from the pressures that buffet the organization. One must hasten to add, however, that the distinction between the freedom of the RAND researcher and the freedom of the organization is far from a perfect one. There are obviously instances where the two are inseparably linked. Although it may sound paradoxical in light of the foregoing reasoning, one might even observe that a certain amount of self-restraint and self-control on the part of RAND researchers could contribute to keeping the organization free.

RAND seems to face a future that is both cloudy and bright. On the one hand, its opportunities for service in many ways have never been greater. But at the same time problems have emerged in more acute forms than RAND has ever faced in the past, and it is clear that the organization cannot continue to exist simply on the basis of its past reputation.

---

[59] RAND also has what is known as the "20 Percent Policy" that applies to all senior research personnel: each researcher is entitled (and expected) to spend 20 percent of his time keeping abreast of developments at the frontiers of knowledge in his field. (IDA has a similar policy, and most of the other nonprofit advisors pay some sort of obeisance to the principle.) In recent years, however, there has been a tendency for top management and some department heads to interpret the "20 Percent Policy" more strictly and to insist on a somewhat higher standard of relevance for work that is not directly related to current RAND projects.

Many RAND people are genuinely worried that, just at the point when most is expected from RAND, it will be able to deliver the least. Whether or not this will prove true in the years ahead is a major challenge facing RAND and its government clients.

INTERNAL ORGANIZATION

In its early days, RAND was governed by very informal and easy-going administrative practices. The organizational structure was very fluid and adaptable. As the organization grew larger and acquired various new research responsibilities, there arose a greater need for more formal management practices. The RAND Corporation is currently an organization of some 1,100 employees (half of whom are professional) with a complex internal structure; and RAND management is conscious of a continuing need to strengthen and improve RAND's administration. The general aim of this chapter is to describe the salient features of RAND's internal organization, and to analyze some of the main administrative problems that the corporation faces.

THE DEPARTMENTS

As Figure 2 indicates, RAND is subdivided into eleven technical departments: aero-astronautics, computer sciences, cost analysis, economics, electronics, logistics, mathematics, physics, planetary sciences, social science, and system operations. These technical departments, it will be noted, are primarily organized around professional skills rather than functional concepts like strategic planning, space weapons, limited war, and the like. This has two principal advantages. First, it facilitates recruitment of the professional staff. Top manage-

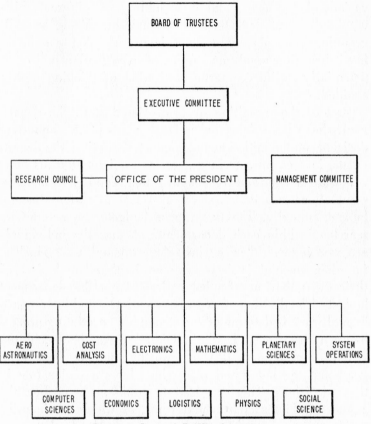

Figure 2. Internal RAND Organization

ment has sensibly delegated the function of hiring the re-search staff to the individual departments. The departments resemble their academic counterparts, and it is generally easier to attract people when they are assured of an opportunity to interact with colleagues in their disciplines. However, it should be noted that not all economists at RAND, for example, are located within the Economics Department. There is some movement of research personnel between various departments, and a sprinkling of different skills within different departments.[1] Second, organizing by skill groupings prevents any sector of the organization from thinking it has a monopoly on studies of a particular subject. A potentially sterile compartmentalization of thinking is thus avoided.

Besides the recruiting of professional staff, the departments have been delegated a large share of the important decisions on the initiation and conduct of studies. The departments, in fact, operate in general as semi-autonomous units with wide latitude and freedom of action. Top RAND management allocates a certain level of funds to a departmental budget annually. The department budgets support the researchers within each department, whether the individuals are working on inter- or intra-departmental research. The interdepartmental projects have no independent budget of their own; they are funded entirely from the department budgets. The division of effort between intra- and interdepartmental work is determined by a complex negotiating process between department head, individual researchers, and top RAND management. It is noteworthy that the department head and the individual researchers have a major role in these decisions. This can lead to problems regarding the support for interdepartmental (most of which are interdisciplinary) projects. The departments may be loath to surrender staff for the support of work outside their own immediate areas of interest. RAND's front office must then

---

[1] RAND has found that a certain intermingling of skills even within the departments broadens intellectual horizons and facilitates interdisciplinary research. An interesting oddity is that several RAND technical departments have philosophers on their staffs. For a discussion of some of the problems of organizing by function, project, or skill group, see Herbert A. Shepard, "Nine Dilemmas in Industrial Research," in Bernard Barber and Walter Hirsch, *The Sociology of Science* (Free Press, Glencoe, 1962), pp. 344-355.

negotiate with department heads to pry loose staff to support interdepartmental projects.

The important point to note for present purposes, however, is the high degree of decentralization that characterizes RAND's internal organization. Indeed, decentralization is the most striking single characteristic of RAND's internal organization.

A high degree of decentralization also prevails within the departments themselves. Individual researchers generally have wide latitude to work on subjects that interest them, to formulate the issues they think significant, and within broad limits to set their own deadlines and work on congenial schedules. Moreover, since department members are normally in fairly close daily contact, there is a considerable amount of informal consultation on departmental matters. Individual staff members often have an opportunity to make their views heard and to participate in the making of departmental decisions.

There are, however, differences among the various departments. Some departments are more hierarchically organized and more tightly directed than others. This results to a large extent simply from the personality and style of the various department heads, but also to some extent from the nature of the department's work. Some departments are more concerned with basic research, whereas others tend to be more oriented toward the applied end of the research continuum. Some departments also are to a large extent insulated from immediate client requests, while others are more responsive to requests for short-run assistance from clients. In the latter case, as one might anticipate, the internal department organization tends to be tighter and to reflect a more explicit direction of the work effort by the department head. In all RAND departments, however, there is a standing policy of

151

encouraging individual researchers to spend some portion of their time in research and study of their own choosing designed to broaden their professional competences.[2]

THE MANAGEMENT COMMITTEE AND THE RESEARCH COUNCIL

The departments also share in the making of the corporation's broad management decisions through the device of the Management Committee. The Management Committee, which meets weekly for several hours to discuss corporate policies and issues, consists of the eleven department heads plus RAND President Collbohm and Vice Presidents Goldstein and Henderson. The agenda of the typical Management Committee meeting covers a wide variety of topics, ranging from small administrative matters to discussions of such basic questions as orientation of the research program and the pros and cons of taking on a new contract. It should be noted, however, that the Management Committee is *not* a decision-making body. It is an advisory body only, and its meetings are discussion sessions which are used by the RAND President in arriving at decisions. Final decision-making authority remains firmly lodged in the Office of the President, subject to the broad policy guidance of the RAND Board of Trustees.

A second advisory body to RAND's President is the Research Council (whose creation will be discussed below in connection with the RAND reorganization of 1960). The Research Council consists at present of five senior RAND researchers plus two associate members, who are also senior research people. Its essential task is to perform a planning function and to make recommendations to RAND's President on measures needed to coordinate the research program. In practice, it has come to act primarily as a council of elders who offer advice when requested to all parts of the organiza-

[2] See Chapter IV, note 59.

tion but who have no formal managerial responsibilities. The Research Council has no staff, its meetings are infrequent and very informal, and its members typically devote much of their time to pursuing their own research interests.

The decision-making authority of RAND's President extends in theory to all aspects of the corporation's internal affairs, but in practice there are a number of constraints set by custom and tradition on the exercise of this authority. For example, except in unusual circumstances top management does not interfere with department privileges in regard to the hiring (or firing) of the research staff. Nor is it likely that RAND's front office would exert pressure on the departments to assign certain tasks to certain personnel. In brief, the unwritten rules of RAND's internal administration lay down a fairly well-defined sphere of department autonomy. Another sphere represents a sort of gray area where the departments and top RAND management engage in negotiations, consultations, informal give-and-take. The decisions which emerge here could be considered joint decisions. Examples from this broad and less well-defined sphere include salary review procedures and issues relating to the execution of interdepartmental projects. In the area of broad management issues there has never been any question that decisions will be made by the Office of the President.

As in all research organizations, of course, there are elements of tension and conflict between the research staff and the administrative staff. This conflict is reflected, in RAND's case, in the departments seeking to widen the sphere in which their influence is predominant and the central management staff tugging in the opposite direction toward greater control of department activities. It is also reflected within the departments in disputes between the department head and researcher over, say, the amount of time some staff member

153

may spend on RAND-sponsored as opposed to client-oriented research, the extent of an individual's outside commitments for teaching or writing, the number of supporting staff a researcher can command for his own research interests, and numerous other matters. Some such conflict between the administrator and the researcher is probably inevitable in all research organizations, and should not be a cause for alarm. Kept within certain bounds, conflict can act as a healthy stimulus to an organization.[3] Management-staff conflict is generally tolerable in an organization like RAND so long as a reasonable correspondence is assumed to exist between the interests of the organization as a whole and the professional self-interest of the research staff. Such conflict would become seriously disruptive if the abstraction "RAND" were taken too seriously, and invested with a corporate personality and identity of its own with discernible interests which consistently ran counter to the researchers' professional interests.

THE MANAGEMENT PHILOSOPHY OF FRANKLIN R. COLLBOHM

Clearly, the personality and style of the President is likely to leave a strong mark on RAND's method of operation and internal organization. It is worthwhile, therefore, to examine briefly the "management philosophy" of Franklin R. Collbohm, who has been RAND's President (or Director) since the organization's inception. Collbohm is an example of that

---

[3] As Coser observes, "Groups require disharmony as well as harmony, dissociation as well as association; and conflicts within them are by no means altogether disruptive factors. Group formation is the result of both types of processes. . . . Far from being necessarily dysfunctional, a certain degree of conflict is an essential element in group formation and the persistence of group life." Lewis A. Coser, *The Functions of Social Conflict* (Free Press, Glencoe, 1956), p. 31. See also Melville Dalton, "Conflicts Between Staff and Line Managerial Officers," *American Sociological Review*, 15:342-351 (1950), and Rensis Likert, "A Motivational Approach to a Modified Theory of Organization and Management," in Mason Haire, ed., *Modern Organization Theory* (John Wiley, New York, 1959), pp. 184-217.

new species in American life ushered in with the age of technology—the research administrator. Though not a research scientist himself, Collbohm has spent most of his career managing scientific and technological activities. Trained originally as an engineer, he recognized the potential of aviation at a time when the industry was in its infancy. He joined the Douglas Aircraft Company where he became one of the nation's first test pilots. Subsequently he became special assistant to Arthur E. Raymond, Douglas Company Vice President and Head of Engineering. During World War II Collbohm participated in one of the original applications of operations-research techniques to military planning.[4] At the war's end, he belonged to the circle of individuals that sought to convince the military of the revolutionary implications of organized scientific research for the future military establishment. His background fitted him well for the role of Project RAND Director. From the start Collbohm sought to operate RAND by surrounding himself with men in whom he had confidence and giving them wide discretion to do what they considered interesting and significant. He had had considerable experience in dealing with the eccentricities of creative researchers and was wise enough not to impose any rigid framework of inquiry on the staff. He intervened as little as was consistent with his maintaining an over-all management control of the project. Initially, also, his contacts with the research staff were very informal. Many decisions were taken by simply chatting with people and reaching a consensus on what should be done. Although these procedures have had to become considerably more formal over the years, strong traces of Collbohm's initial approach remain in RAND's internal administration.

[4] See Chapter II.

Despite some modifications, Collbohm's management philosophy also remains essentially the same. He places continued stress on decentralization as a desirable means of making decisions on the research program: the individual researcher generally knows best what should be studied and how it should be studied. Delegation of authority is generally desirable within certain broad limits. Overadministration can often be more dangerous than inadequate central administration. Such central management control as is necessary should normally be of a general nature; it should concentrate less on specific details than on over-all performance. However, some specific controls have been considered necessary. One area, for example, which Collbohm has always believed of special concern to management is the matter of salary review and salary raises. No department head can unilaterally raise a man's salary; such action has to be cleared with top RAND management through the mechanism of the Salary Review Board. This board consists of Collbohm, Goldstein, a member of the Research Council, and the relevant department head. The board jointly reviews semiannually the salary position of every professional staff member in the various RAND departments. With the number of professional employees in RAND now numbering over 500, it can be seen that this salary review procedure represents a considerable effort and consumes a significant amount of top management's time.

In sum, Franklin R. Collbohm's management philosophy has in general placed high stress on research staff initiative and creative self-expression. This broadly permissive approach to the leadership of RAND has undoubtedly contributed much to its healthy research atmosphere. Remarks like the following from senior RAND research people attest the generally permissive atmosphere one finds at RAND:

At the time of my arrival [1949] I was impressed by the anarchy of both policy and administration which I found in Santa Monica. Since such an atmosphere exists only rarely . . . I was shocked and disturbed . . . [But] I soon was carried into a stream of interests in which questions of orderliness in corporate policy and administration no longer seemed important. Subsequently, I have come to regard this anarchy as the strength if not the substance of RAND . . . Obviously, what I am calling anarchy is not really anarchy but rather a degree of intellectual freedom which is so unique that it seems like anarchy in terms of my earlier experience [in government and academe].[5]

THE RESEARCH PROGRAM

To speak of a RAND research "program" is actually somewhat misleading. In the sense of a coherent objective, with all parts of the research effort integrated nicely within a common framework, there is no research "program" at RAND. It would be more accurate to speak in terms of a research "menu"[6]—a varied collection of items reflecting the ideas and tastes of many cooks. There never developed at RAND any particular research "mission" other than the very broad and amorphous one of studying "preferred techniques and instrumentalities of intercontinental warfare." Some of RAND's founders and early backers preferred to see RAND's efforts focused around a specific objective—for example, a research and development program leading toward the development of a ballistic missile. But it probably has been fortunate for RAND that no explicit program approach was followed initially and that the research effort was able to develop along a number of different lines. The variegated "menu" is probably more deeply rational than any superficially coordinated and germane program.

5 Statement by David Novick, head of the RAND Cost Analysis Department, from an internal memorandum, June 19, 1961.

6 I am indebted to the late John D. Williams for this formulation, which appears in an internal paper, "An Overview of RAND," D-10053, May 14, 1962.

As a result, however, we are left in something of a quandary in attempting to answer the question: what kind of research does RAND do? The unsatisfactory short answer is that RAND does a bewildering variety of different research assignments. Short of an encyclopedic listing of some of the various items in RAND's unclassified index of publications,[7] it is difficult to convey an impression of the range and breadth of the RAND research "menu." Yet even such a ponderous approach would be inadequate and misleading. For the security problem presents an additional obstacle in portraying the full scope of the research effort. The unclassified portion represents only a part of the total effort with some of the most important research being done on a classified basis.[8]

Perhaps the best approach to a description of the RAND research effort that has the advantages of both conciseness and substantial accuracy would be to divide the effort into three broad mutually-supporting categories: (1) basic research, (2) systems analysis, and (3) miscellaneous research assignments. First, it has been a widely-accepted principle at RAND that some portion of the total research effort should be devoted to "pure" or undirected research within the traditional disciplines. This is necessary to help provide the basic intellectual capital for interdisciplinary projects, and to

[7] See the Index to Unclassified RAND Publications, available in annual editions at the RAND Deposit Libraries across the country. The index is a lengthy volume of some 700 pages.

[8] At the author's request, a RAND office undertook a random sample check of the index to all RAND studies to obtain a percentage breakdown between classified and unclassified studies. A check of four different time periods before 1958 revealed a 50 percent to 50 percent breakdown (number of studies in sample: 300). A more intensive check of the period January-May 1962 showed a breakdown of 54 percent unclassified to 46 percent classified for RAND "D's" (Documents), and 64 percent unclassified to 36 percent classified for all Reports, RAND Memoranda, Papers, and Documents. Number in sample: 793. From this one can obtain a rough inference about the approximate division of effort in RAND between classified and unclassified work.

create the kind of academic atmosphere required to attract talented researchers.[9] Thus all the RAND departments engage in some pure or basic research. For example, the Computer Sciences Department, which performs primarily a service function for the other RAND departments, engages in basic work on the mathematical theory underlying computer uses, computer simulation of human thought processes, mathematical biology, computer applications in the study of linguistics, and other areas.

The point must be carefully stated. Saying that RAND departments all carry on some basic research may mean that researchers in a department divide their time between more basic and more applied work. Or it may mean that some researchers do essentially basic work all the time, while others perform essentially applied work full-time. There is a tremendous variation in the professional experiences of researchers at RAND. Some, especially those with established reputations, pursue their scholarly interests without regard for the work's "relevance" or for client concerns. Others are expected to produce from time to time work that is of interest to one RAND client or another. Some researchers need direction, others resent it. An individual's experience is shaped by his background, abilities, market value, relationship with the department head, and the expectations on which he was hired. Departmental differences are also evident.

---

[9] This consideration may be relevant to the experience of the developing nations as they face the task of building research institutions for various needs. It seems unlikely that a new nation can merely create "applied" research institutions concerned only with transmitting and applying technical knowledge gained elsewhere. An effective research capability requires an indigenous and on-going effort, and an organizational setting with some individuals engaged in research at the frontiers of knowledge. Without the latter it may be difficult to attract and hold a competent research team. See Edward Shils, "Scientific Development in the New States," in Ruth Gruber, ed., *Science and the New Nations* (Basic Books, New York, 1961), pp. 217-226.

Some departments, such as RAND's Planetary Sciences Department, are mainly engaged in basic research, whereas other departments, such as the Systems Operations Department, are more concerned with applied work.

It should be observed here that RAND's ability to carry on some self-sponsored research is an important guarantee that basic research will not be slighted. If basic work cannot be charged off as costs incurred under a contract, the availability of fee income makes it possible for such work to continue as RAND-sponsored research. In general, however, RAND has had little difficulty persuading sponsors that some basic work is a legitimate part of the research enterprise.

Second, a large and significant part of the RAND research effort consists of analytical work best described by the term "systems analysis." As previously suggested, systems analysis developed out of earlier operations-research techniques and represents a more sophisticated approach to the increasingly complex military problems of the postwar era. It is here that we find the interdisciplinary focus of RAND research; and it is here also that much work of direct relevance to client concerns is done.

The concept of systems analysis is difficult to define precisely.[10] The term "system" generally came into use to denote an effort to be more comprehensive than the traditional operations-research study. Systems analysis sought to consider more than the use of particular weapons in a specific operational context, and generally attempted to take into account all the relevant factors affecting a complex problem

[10] See Malcolm W. Hoag, "What Is a System?" *Journal of the Operations Research Society of America* 5 (June 1957); James R. Schlesinger, "Quantitative Analysis and National Security," *World Politics* 15:295-315 (January 1963), E. S. Quade, ed., *Analysis for Military Decisions*, The RAND Corporation, R-387 (November 1964); and N. Jordan, "Some Thinking About 'System,'" The RAND Corporation, P-2166 (December 30, 1960).

under investigation. A "system" in this context can mean a new weapons system and all the interrelated economic and strategic considerations associated with its development. Thus RAND has, for example, carried out systems analyses of the B-70, the Minuteman missile, the Midas reconnaissance satellite programs, and a modest level of effort assessing the capabilities of the Polaris missile-firing submarine. Or "system" can refer to the broader and more elusive concept of a total constellation of military forces and political constraints that may govern the implementation of a broad strategic objective. Thus RAND has done broad studies on limited war and on the role of air power in limited-war situations in various parts of the world. Such studies have involved consideration of weapons capable of use from aircraft, supply costs of different types of operations under various assumptions of combat, problems of coordinating operations with allies, and a host of other intricate issues. Finally, "system" can suggest something much more limited than either of the above definitions. For the purposes of a particular analysis, a "system" can mean a subsystem or component part of a total weapons system or force posture whose effectiveness is compared to some other subsystem.

It is clear that there are systems within systems: reality does not divide itself neatly into self-evident categories for the analyst's convenience. Consequently, like any intellectual endeavor, the systems study must take great care to define the parameters of the subject it proposes to analyze. What is assumed as "given" and outside the scope of one study may be the critical system variable analyzed in another. Obviously, the realism of the assumptions and the care taken to define the relevant variables crucially affect the study's quality and usefulness. In general the more specific the "system" to be studied, the easier it will be to arrive at clear-cut solutions.

But this type of approach has limited utility on the broad policy questions where objectives cannot be precisely specified and no "optimum" solutions are possible.[11] The useful systems analysis in this area typically attempts to identify relevant policy alternatives under conditions of great uncertainty, and helps the policy maker to assess the costs and consequences of different courses of action. The RAND systems analysis over the years has generally moved in the latter direction. It has grown less rigorously mathematical in nature and more concerned with assessing the large uncertainties and qualitative factors that affect decision on an important problem. However, a broad systems analysis may still contain as subelements—or "inputs"—to the analysis a variety of quantitative research techniques. Indeed, a useful systems analysis may contain so many varied "inputs" that one sometimes has the feeling this type of research effort can be summed up with the injunction: Be as systematic as possible in applying a variety of tools to an analysis of a complex problem. In the following chapter a significant RAND systems analysis is examined in detail to give the reader a better

[11] This is one of the crucial distinctions between decision making in the firm and in the government agency. In the case of the firm, there is a clear objective (profit) that can be measured in unambiguous units of analysis (dollars). And the basic market environment in which decisions must be made can often be assumed as fixed and unaffected by management actions. Hence, precise optimizing techniques can provide ready guides to decision on a wide range of management problems, especially at the middle and lower management levels. In contrast, the "political" decision involves multiple objectives that are unclear or in dispute, and the outcome of each decision affects power alignments and so influences the context in which new decisions must be taken. See Robert A. Dahl and Charles E. Lindblom, *Politics, Economics, and Welfare* (Harper, New York, 1953), esp. chs. iii, xiii and xiv, Harry Eckstein, *The English Health Service* (Harvard University Press, Cambridge, Mass., 1958), ch. ix, and Albert Wohlstetter, "Strategy and the Natural Scientists," in Robert Gilpin and Christopher Wright, eds., *Scientists and National Policy-Making* (Columbia University Press, New York, 1964), pp. 174-239 and "Analysis and Design of Conflict Systems," in E. S. Quade, ed., *Analysis for Military Decisions*, pp. 103-148.

understanding of what such a study is and how it is carried out.

Third, RAND performs a number of miscellaneous research and advisory tasks that do not fit neatly into either of the above categories. For example, RAND may loan several staff members temporarily to a government agency to assist with some pressing project,[12] comment informally on certain contractor proposals submitted to a RAND sponsor, carry out "quick and dirty" studies on matters where decision cannot be delayed, supply technical information to a sponsor or comment on studies in progress, and send RAND people to Washington or field installations for informal discussions with government officials. RAND people are not always anxious to fulfill such requests. But since RAND does have a service obligation to its clients and can occasionally be quite useful through such activities, it is widely felt that such requests should be honored whenever feasible. It is also true that such activity, if taken in moderation, can be quite useful intellectually to RAND by introducing new points of view and suggesting new problems for study.

Little purpose would be served in attempting to achieve a spurious exactness by assigning percentage figures to the three general categories of RAND research. It is impossible to obtain any meaningful statistics for the total man-hours of research effort devoted in any given time period to basic research, systems analysis, or miscellaneous research activities. The various research activities, in any case, are interrelated and to a considerable extent mutually supporting. Systems

12 This most extensive operation of this kind was the assistance RAND provided the Office of the Secretary of Defense/Comptroller in establishing the program budget system in the Defense Department. A special RAND office was created in Bethesda, Maryland for this purpose, and some 15-20 RAND professionals were involved full-time in the activity during the 1961-1963 period.

analyses draw on basic research, and basic work can evolve out of an insight gained in a systems study or even in a "quick and dirty" project. About all that can be said on the point is that the bulk of the RAND effort generally revolves around systems-analysis work, but basic research and various miscellaneous research assignments also figure importantly in the total effort.

It may be of interest, however, to mention the approximate allocation of effort among the various RAND sponsors. A break-out based on the percentage of the research effort charged to each contract or grant category (or to RAND-sponsored research) reveals the following distribution:[13] Air Force (Project RAND plus other Air Force work), 68 percent; Department of Defense, 17 percent; National Aeronautics and Space Administration, 6 percent; Atomic Energy Commission, 2 percent; RAND-sponsored research, 2 percent; Agency for International Development, 2 percent; National Institutes of Health, 1 percent; other, 2 percent.

This distribution, however, suggests an air of greater precision and compartmentalization than actually obtains in the conduct of the research program. The individual RAND researcher may sometimes lack a clear idea of what percentage of his time is going to what contract. Occasionally, it may happen that an individual will be doing research in one general area and yet have his time divided up for administrative purposes into NASA work, Project RAND work, and OSD work. To some extent, therefore, the precise allocation of effort among the different sponsors amounts to a convenient administrative fiction which may have little relevance for the individual researcher. Yet it is not only a fiction because RAND would not undertake work for a client unless

[13] Compiled from internal documents obtained from RAND official M. C. Davie.

some RAND researchers were definitely interested in working on the subject-matter area covered in the contract agreement with the client. As RAND moves toward a larger number of clients, and faces the prospect of having numerous contracts and more explicitly defined obligations, there will probably be an increasing tendency for researchers to be specifically identified with particular contracts and particular clients.

As a last phase of the research effort, RAND also supports from its own funds a small program of student fellowships awarded to graduate students in various educational institutions. In 1965, eight such fellowships were awarded in the fields of mathematics, physics, aeronautics, engineering, systems and communications sciences, economics, and international relations to such institutions as Harvard, Yale, Princeton, the University of California, the University of Chicago, Stanford, and Columbia University.

INCEPTION OF RAND STUDIES

How do RAND studies get started? In response to this question, we are again forced to paint a somewhat fuzzy picture. A certain amount of fuzz, indeed, seems to be an essential attribute of most phases of RAND's administration. A chief initial difficulty arises in attempting to assess the extent to which RAND studies are undertaken in response to client requests and to what extent they are self-initiated. This is obviously a crucial issue, for the research organization that only responds to client initiatives may lose the capacity for detached work.

A fairly accurate general statement is that many RAND studies originate from a *consensus* between the client and the research organization. That is, through informal and formal negotiation, substantial agreement is reached on what should be studied, reflecting both the client's interests and RAND's own assessment of what research it can and should do. That this consensus concept has some genuine content is shown by the fact that RAND reserves the right not to undertake particular studies. In actuality, RAND refuses a considerable number of requests for studies each year even from the Air Force, its oldest and most important sponsor. Yet there are certainly limits to RAND's ability to stand aloof. If the Air Force Chief of Staff approaches RAND with a request for a study he considers urgent, for example, the odds are high that RAND will find merit in the chief's suggestion and get some people working on the study. Thus, while a consensus generally prevails, there is apt to be a certain fluid and dynamic element to the consensus. Sometimes the client interest appears as the dominant influence in the inception of studies, and sometimes the RAND influence is predominant. (This statement, of course, does not apply to RAND-sponsored research projects, which are wholly RAND-initiated.)

The further qualification must be entered that some variation exists with regard to different sponsors; some RAND sponsors are more insistent than others on having a hand in the initiation of studies and the direction of the research effort. Work under the Project RAND program, for example, consists to a substantial degree of a large number of small studies that are frequently initiated by the RAND researchers themselves. ARPA, ISA, and NASA work, on the other hand, and the larger studies done under Project RAND, reflect a greater degree of client participation.

Several years ago the Air Force Project RAND Liaison Officer made an effort to identify precisely what percentage of RAND studies resulted from client requests and what percentage was self-initiated. He found for the fiscal year 1959 that the following percentage of study items were self-initiated or initiated by client request:[14] 27 percent requested formally by the Air Force RAND Liaison Office; 6 percent requested informally by the Air Force; 18 percent initiated by RAND, but with the direction of the effort changed significantly by the Air Force through suggestions, modifications, and changes; 5 percent requested by other Air Force commands, such as the Air Research and Development Command (now Systems Command); 37 percent initiated by RAND with no Air Force influence; 7 percent work for sponsors other than the Air Force.

This effort is interesting mainly for the general light it throws on the interaction between the client and RAND in the orientation of the research effort. The precise percentage figures, besides being rather dated, cannot be taken very seriously. There is no clear way of making such precise distinctions, and the figures therefore have a spurious exactness. But at least they serve to indicate that a portion of RAND studies tend to be self-initiated, another portion rather more client-initiated, and a third portion of mixed inception with the result that it is difficult to determine where the major initiative rests. Further, there is no clear dividing line. The three groupings form a continuum and shade off gradually into different degrees of client influence or RAND influence in the selection of studies.

Within RAND itself the initiative for undertaking studies comes to a considerable extent from the research staff. Top management seldom attempts to direct the departments to perform particular studies but occasionally RAND's President or Vice President will informally suggest a topic of

[14] *Department of Defense Appropriations for Fiscal Year 1961*, Hearings before a Subcommittee of the Committee on Appropriations, U.S. House of Representatives, 86th Cong., 2nd Sess., part 7, p. 184.

inquiry. And sometimes a study is initiated in this way. Sometimes a department head will suggest a study to his staff, while on other occasions the impetus for a study will come from an individual staff member. At all levels, from central administration to department head to individual researcher, there are various formal and informal contacts between RAND people and government officials so that client suggestions and ideas are fed into the internal RAND decision-making stream at a number of places.

A slightly different pattern prevails when one considers the actual execution of a study once it has been initiated. Here the client influence tends to be diminished. RAND jealously guards its freedom of action to define operationally how a problem will be approached and how in general the research will proceed. In this respect, RAND differs significantly from some of the other analytic advisory institutions that work more closely with the client in the planning and actual execution of research projects. In WSEG and Hum-RRO, for example, half of the research personnel are from the military. A dual military-civilian chain of command exists which makes joint decisions not only on the inception of studies but also on how studies are to be carried out. RAND, in contrast, makes its own "in-house" decisions, has no military administrator with responsibility for the corporation's operations, and does not have a sizable group of military personnel within the organization working directly in the research effort. RAND does have, however, ten Air Force officers assigned to full-time duty at RAND. But these men have no operational responsibilities or control over the research effort. They act either as observers or they may join in RAND studies as part of the normal research staff subject to the supervision of their department head and central RAND management.

There remain, nevertheless, elements of cooperation and consultation with clients during the research and writing phases of a study. Many RAND studies, for example, require information from client agencies for their successful completion. This is one important source of continuous contact with the client. Further, RAND people may provide the client with an informal progress report on a study and solicit comments and suggestions. Such a technique can be particularly useful when it comes to communicating the research results to the client. Officials of the client agency tend to feel they have participated in the study and, therefore, may be more receptive to the study's policy recommendations.

There are a number of subtle questions in deciding how far the advisory organization should go toward soliciting client participation in a research project. On the positive side, the policy of keeping somewhat aloof from the client is probably a basic safeguard against client domination. Yet, at the same time, there is a price to pay for insulating research operations from significant client participation. In particular, there will likely be increased difficulties in "selling" the research results to the client.

In recent years, there has been a tendency at RAND to increase somewhat the amount of consultation with the client during the course of a study's preparation. The Air Force officers on assignment to RAND also seem to be playing a somewhat more active role in serving as a communications link between the Air Force and RAND. One indication of this is the fact that several Air Force officers assumed the role of project leaders for important RAND studies. The officers participated directly in the research, and assumed primary responsibility for communicating the research results to Air Force Headquarters.[15]

[15] The two officers were Col. Ross Blachly and Col. William Jones. Blachly's

Such developments may be in part attributable to the operation of "cooptation" mechanisms similar to what Selznick observed in his classic study of the Tennessee Valley Authority.[16] RAND's efforts toward increased consultation with the client seem to have occurred at about the time significant tensions developed in the client-advisor relationship and RAND's external environment in general became more hostile. Faced with possible threats to its existence, an organization may understandably seek to solidify its position by "coopting" actual or potential critics into some role in the organization's activities.

However, as Selznick observed in the case of the TVA, this course of action is likely to have some undesirable consequences. Giving outside elements a role in the organization's actions may result in drastic changes in organizational objectives and operating procedures. Some of these dangers were vividly illustrated in RAND's one experience with extensive client participation in the performance of a study—the Strategic Offensive Forces Study (SOFS) in 1959. This project had unhappy consequences for all concerned. For a variety of reasons, not alone civil-military tensions, the SOFS project floundered badly and turned out to be one of the least successful studies RAND has ever undertaken.

A military officer and a senior RAND researcher were

project dealt with the acquisition of a Chinese Communist nuclear capability, and was carried out largely in 1961 and 1962. Jones headed a project dealing with command and control problems and SAC mission planning that extended over a period of several years beginning in 1956. It should be stressed, however, that the work of these officers does not necessarily form a pattern that other Air Force officers on assignment to RAND will follow. The role played by the Air Force officers assigned to RAND varies greatly and depends importantly on the individual's interests, abilities, and relationship to department colleagues.

[16] Philip Selznick, *TVA and the Grass Roots* (University of California Press, Berkeley, 1949).

named over-all supervisors of SOFS; and the study was then broken down into a number of subprojects. A military officer along with a civilian counterpart from the RAND research staff assumed joint responsibility for directing each of the subprojects. Dozens of Air Force officers flocked to RAND headquarters in Santa Monica during the course of the study. Many RAND people felt the organization was being invaded by the military. There was also a rigid deadline imposed on the study which further irritated some RAND people. The ubiquitous presence of the military, the pressure of the deadline, and the massive and unwieldy size of the project all combined to bog the effort down in a maze of confusion and frustration. (The SOFS project was the largest in RAND's history, involving at one point over 100 members of the research staff.) No final report was ever issued from the study, and a number of RAND researchers reportedly resigned in direct consequence of resentments generated in the course of the effort. The SOFS project would provide an interesting case study in the pathology of research administration. This experience with extensive client participation in actual research operations left a strong residue of opposition in RAND to further experiments along this line. Efforts at improved consultation with clients will probably continue, but it is doubtful that they will resemble the extensive client participation in actual research operations manifested in the SOFS project. RAND seems willing to pay a certain price to maintain essential control over the conduct of the research effort.

DEPARTMENT VS. PROJECT RESEARCH

RAND, we have noticed, is organized around semi-autonomous departments each with research interests of its own and each also having responsibility for participation in interdepartmental projects. This dual responsibility poses some intricate problems. The necessity for some sort of project organization seems to be a perennial problem for all the advisory corporations. Whatever the internal organization—whether along the lines of academic departments or along the lines of functional concepts like limited war, strategic forces, and so on—there seems to be an inescapable need to superimpose upon the structure some sort of administrative arrangement that assigns responsibility for particular projects. Organizing on a project basis alone is one possible answer, but this has decided disadvantages. If project responsibility were the sole basis of internal organization, it would be difficult to develop any consistent pattern of support services for the research staff. The organization would be in a constant state of flux as old projects were completed and new projects emerged, and all administrative decisions would have to be taken on an *ad hoc* basis. Recruitment problems would also be multiplied as the researchers would have no "home" from which to operate. Further, it might become more difficult to drop a project that no longer looked promising—a frequent necessity since good projects are seldom formulated without a considerable amount of preliminary groping and some false starts. It is easier to abandon an unpromising project if researchers know they have a firm place within the organization to which they can return.

RAND has sought to cope with the problem by trying to achieve a balance between department and project research. The different departments and their research staffs are encouraged to engage in research within their own areas of interest. The staff keep up with the literature in their fields, attend meetings of professional societies, and discuss economics or physics or electronics with each other, and engage in incidental professional activities like writing book reviews.

Exploratory research is also encouraged which is not considered part of any project (but which may lead to a project). Then, in addition, the departments and their staffs are encouraged to "interact" with other departments and researchers of different skills and backgrounds, and to take initiatives in organizing interdisciplinary study terms. Interestingly, some of the chief frictions and obstacles to interdisciplinary cooperation arise not between widely distinct skills but between skills that are fairly closely related where, perhaps, each skill group feels a greater need to assert its professional identity.

A "project" formally comes into existence when research has at least progressed to the stage that the problem under investigation is fairly well-defined and when one or more persons have been devoting a sizable portion of their time to the problem for a sustained period. A project number is then assigned to the study, and a project leader is designated. The project leader has the responsibility for the completion of the study, and normally ends up doing a large share of the work. The project is completed when the study finally crystallizes into a report or memorandum that is sent to the client. The project may, however, be dissolved and no completed study ever appear (or only an internal working paper may result). The project can be large or it can be very small, sometimes involving only one or two researchers. The projects are recorded, and their progress checked, by an Office of Program Analysis located within RAND's Office of the President. This office, however, does not "direct" the work. Responsibility for the direction of projects rests with the departments and the project leader. At present there are some 240 projects in progress at RAND.

The lot of the project leader is in many ways an unenviable one. To illustrate the point, let us briefly consider the case of a project leader in a typical RAND systems analysis. Once designated, the project leader first sets out to assemble a team (if the project is at all a sizable one) to assist him in the research. This is by no means easy, for while he has full responsibility he has little or no authority. He cannot command anyone to work on the project. His chief means of obtaining assistance is to make the project seem so exciting and significant that colleagues will be persuaded to join the project team. The project leader thus might be said to have an unrestricted hunting license, but no ammunition. It is not unknown for projects never to get underway for lack of adequate staffing. Then it becomes necessary to appeal to top RAND management and/or to the heads of other departments for assistance. A complex negotiating process will then take place whereby staff is pried loose from departments to support interdisciplinary projects.

Once he has assembled his team, the project leader's troubles are by no means over. Holding intact a group of diverse researchers is itself a substantial achievement. But getting different talents to work together creatively is a high art.[17] Research results do not emerge automatically from group interaction, and the project leader thus discovers that he must *lead*. His leadership, however, must not be so aggressive as to stir resentment among the team members. The usefulness of some RAND people as project leaders has been vitiated by real or fancied "bullying" attributed to them during previous service as project leaders.

If the team happens to be large, the project leader's problems are apt to be more difficult. Luszki points out that

---

[17] A comprehensive treatment of this topic is Margaret Barron Luszki, *Interdisciplinary Team Research: Methods and Problems*, no. 3 of the Research Training Series, published for the National Training Laboratories (New York University Press, New York, 1958).

personal differences between team members and project leaders (and among the team members themselves) are likely to increase as the size of the project increases.[18] Furthermore, as teams become larger, more time is needed for administration and communication to keep the members of the team aware of what the others are doing.[19] The project leader may be forced into an administrative role that he is not well qualified to fill and that wastes his research abilities.[20]

Meanwhile, the members of the team are apt to be drawn back toward their own departments by a series of informal pressures and inducements. This is likely to be the case particularly if the project extends over a long period of time, and if the members of the project team are devoting part of their time to other projects or research interests. The heads of the departments who have "loaned" staff members to a project may be anxious to have them back for research of more immediate interest to the department. The individual researcher's normal pull is toward his department in any case since his professional standing depends primarily upon his performance within his own discipline and upon the judgment of his peers within the immediate department "subculture." The salary factor, to the extent that this motivates the individual researcher, also operates as an inducement to the individual to be more sensitive to his departmental, rather than to his interdisciplinary, responsibilities. Salary increases are recommended by the department head to the Salary Review Board, not by a project leader who may have had the researcher as a member of a project team.[21]

[18] *Ibid.*, p. 226.
[19] *Ibid.*
[20] *Ibid.*, p. 301.
[21] Project leaders of important studies, however, will often volunteer forceful views to RAND management on salary rewards for persons working on the project. Also, as noted below, the reorganization of 1960 sought to make salary

All of which suggests that the job of project leader is a difficult and demanding task and that interdepartmental project research at RAND is hardly a simple matter. By the end of the 1950's, in fact, many RAND people were seriously worried that interdepartmental work was becoming more and more difficult. A number of internal reforms began to be seriously discussed which would improve RAND's ability to carry out the interdisciplinary projects long considered a vital part of the research effort. Suggestions were put forth, for example, for the creation of a new "floating" department devoted exclusively to interdisciplinary systems studies.[22] Other proposals vigorously debated in RAND included recommendations for budgetary incentives for departments and individuals to devote more time to interdisciplinary work. It was in part due to these pressures, and to cope with the problems of interdepartmental work, that the significant RAND reorganization of 1960 took place.

THE INTERNAL RAND REORGANIZATION OF 1960

RAND has had two noteworthy internal reorganizations—in 1955 and in 1960. Because the more recent and more significant reorganization in a sense represents a reaction against the earlier one, it may be well to begin with a brief description of the 1955 reorganization. RAND by 1955 had begun to reach a size which made it difficult to operate under the very informal management practices typical of the early period. Thus the reorganization of 1955 sought, as one important objective, to formalize RAND's internal structure and to make more explicit assignments of responsibility for different

---

increases more responsive to performance on interdisciplinary projects by providing for Research Council representation on the Salary Review Board.

[22] From a Feburary 22, 1960 memorandum of Albert Wohlstetter to RAND's Management Committee.

aspects of the research program. A second objective of the 1955 reorganization was to eliminate some costly duplication of effort that appeared to exist under the previous fluid structure. This was to be accomplished by centralizing within one large division the functions formerly performed within three separate units responsible, respectively, for missiles, aircraft, and electronics. The argument for centralization was that the separate units—in particular, the missile and aircraft units—were concerned with the same kind of technologies and aerodynamic questions. Therefore it was unnecessary and wasteful to have separate units performing essentially the same function. The resulting change in structure left RAND with three large divisions—engineering, mathematics, and economics—and two smaller divisions—physics and social science.

The latter half of the 1950's was a period of substantial growth for RAND. In addition, it was during this period that RAND began to diversify its sponsorship, which further complicated the management of the research effort. By the end of the 1950's, in connection with some of these developments, the partial centralization effected in 1955 proved to have some serious disadvantages unforeseen at the time of its adoption. For one thing, the engineering division had grown so large—by 1960 it contained some 40 percent of the professional personnel at RAND—that the division head could hardly keep track of the work going on within the division. To some extent, the same problem also applied to the large economics division. However, in the economics case the division head did not really attempt to direct the work of the entire division. The division was something of a loose holding-company for three rather autonomous departments— Economic Analysis, Logistics, and Cost Analysis—and the changes effected by the 1960 reorganization in this case

177

served mainly to formalize an already existing *de facto* arrangement. In any case, a number of RAND people felt that the size of the large divisions, particularly the engineering division, hampered research efficiency and complicated the conduct of interdisciplinary work. By breaking up the divisions formally into autonomous departments, each with representation on the Management Committee, it was thought that the individual departments might be more responsive to RAND-wide problems.

Thus by the end of the 1950's a widespread feeling had grown up within RAND that some kind of organizational changes were necessary. With the unhappy experience of the SOFS project fresh in mind, serious discussion of possible organizational changes began in late 1959. After months of discussion, President Collbohm announced the following organizational changes in a 31 October 1960 memorandum to the RAND staff:[23] (1) the five technical divisions would be broken up to form eleven smaller technical departments each with representation on the Management Committee and standing more or less in the same relationship to RAND's President as the former divisions; and (2) a Research Council composed of six senior members of the research staff would be established to provide general direction and planning of the RAND research effort and, in particular, to assist project leaders in carrying out interdisciplinary projects. To this latter end, it was specified that the Research Council would have salary information and would be represented at all meetings of the Salary Review Board. Presumably, RAND researchers who had performed well on interdisciplinary projects would thus be assured of a fair hearing when salary increases were considered. Further, the Research Council was

[23] Memorandum (M-6147) to RAND Staff from F. R. Collbohm, "Organizational Changes," October 31, 1960.

to have an opportunity to make recommendations to RAND's President concerning department budget allocations. These recommendations would reflect the Research Council's evaluation of department participation in interdisciplinary projects. The Research Council was also to use the prestige of the Council members to assist in getting worthy interdisciplinary projects established and staffed, and generally to be available for advice and consultation whenever problems developed in the execution of such studies.

As Herbert Emmerich pointed out in his valuable essays on the subject, no reorganization is ever final or definitive in any organization.[24] Reorganization is a *continuous* process of adapting administrative forms and procedures to changing needs. Yet, to the extent that we can evaluate the 1960 RAND reorganization at this stage, it would seem that the changes represent an important step toward enhancing RAND's effectiveness and will probably remain in outline the organization's basic administrative structure for some years.

In general, the administrative changes seem to have achieved the objectives they were designed to achieve, although for a time after the organization there was some fear that the less pyramidal organizational structure might actually impede interdisciplinary work.[25] Interdisciplinary work involving several departments seems to have come to function somewhat more smoothly under the new system, and the

[24] Herbert Emmerich, *Essays on Federal Reorganization* (University of Alabama Press, Birmingham, 1950).

[25] The fear arose because L. B. Rumph and Charles J. Hitch, who headed the old engineering and economics divisions, favored interdisciplinary work and used such influence as they had to promote it. The heads of the new departments were somewhat more oriented toward their own specialties, and the more decentralized structure created by the reorganization allowed more scope for the departments to pursue their own separate interests. Gradually, however, the several departments appeared to become conscious of their RAND-wide obligations and the initial alarm felt by some RAND people eased.

179

project leader's position appears to have been strengthened. The breaking up of the large engineering division seems to have increased contact between department head and staff, and has largely removed the image of that division as a huge monolith *vis-à-vis* the rest of RAND. The tendency of the individual departments to be "inward-looking" remains, however, a problem that requires constant attention.

On the perennial issue of decentralization in research administration, the RAND experience seems to reflect a full-circle evolution. Originally, decentralization largely prevailed. Then with the 1955 reorganization a shift toward somewhat greater centralization occurred. Finally the 1960 reorganization moved back toward a more decentralized structure. Notwithstanding, one must caution against drawing any unambiguous "lessons" from the RAND experience in this respect. For RAND has found it necessary since the early 1950's to have some planning effort as an aid to management. At various times RAND has had a Research Planning Committee, a Projects Manager, and now the Research Council entrusted with a general planning role. Despite the unquestioned virtues of decentralization in research administration, it apparently seems essential for any research organization of a certain size to engage in some kind of formal planning of the research effort.

Two incidental effects of the 1960 reorganization deserve brief mention. First, five of the six senior members of the research staff elevated to the Research Council as part of the 1960 reorganization were the heads of the five former divisions. This meant that the way was open for younger men to become department heads and members of the Management Committee—thus providing to some extent an infusion of new vitality into the organization. The move was in keeping with the general RAND tradition of maintaining, as far as practicable, younger men in key research positions. Second, the reorganization served a useful heuristic and morale-building function. The discussion of the proposed organizational changes made people better aware of their problems and contributed to greater mutual understanding among different RAND groups. Furthermore, the fact that many RAND people *believed changes were necessary* made it important for something to be done to fulfill that expectation. The 1960 reorganization may have led to improvements in part simply because staff members acted on a pattern of expectations which assumed that a reorganization would be beneficial—in short, a Hawthorne effect was operative.

It is important, however, not to overstate the impact of the 1960 reorganization. The reorganization worked no miracles, nor did it dramatically transform RAND's internal structure. The success it achieved did not come overnight. Effort, gradual adjustment, and some experimentation were required— and doubtless will continue to be required.[26] Even after the

[26] For several months after the reorganization took effect, the Research Council occupied an anomalous position. It was not clear whether the Research Council was to have a role in management or whether it was to be purely an advisory body. Initially, the Research Council had a staff which attempted to engage in a detailed review of research operations and to exercise a rather ill-defined supervisory responsibility over interdepartmental projects. Jurisdictional clashes arose almost immediately between the Management Committee and the Research Council, and between the Research Council's staff and the staff of RAND's front office. The problem was resolved in the spring of 1961 when the Research Council withdrew entirely from any management role, and transferred its staff to RAND's front office. The Research Council became thereupon much less of a formal structured body: meetings became irregular and less frequent (often simply luncheon conferences), minutes were no longer kept, and council members devoted as much of their time to their own research interests as to council business. The council's influence now depends largely on the prestige of its members and their ability to persuade and cajole colleagues. An area where the council is apt to be particularly influential concerns the selection of project leaders and the staffing and execution of interdisciplinary projects.

Interestingly, the Research Council continues to function even though there

new system began to function relatively smoothly, a number of traditional hard problems have remained which will continue to challenge the imagination and resourcefulness of RAND management. Prominent among these is the task of assuring that RAND will continue to provide a climate for creative interdisciplinary research. The "pancake" type of organization created by the 1960 reorganization could, indeed, pose more problems for interdisciplinary work than the former "pyramidal" structure if the departments ever became dominated by men exclusively interested in their own specialties.

GOVERNING OF THE BOARD OF TRUSTEES

A characteristic of RAND as a private nonprofit corporation is that it operates under the broad managerial direction of a Board of Trustees composed of private citizens. What is the actual role and significance of the Board of Trustees? In what sense does it "govern" or "manage" RAND's activities? Under the terms of RAND's articles of incorporation and by-laws, RAND has 21 members on its Board of Trustees. Three of whom must be officers of the corporation and the others are to be selected, evenly if practicable, from per-

---

is no "consensus" among the members on the appropriate council role. Several members believe that research cannot be "planned" or "organized" and hence have chosen to abstain from these aspects of the council's work. Other members have chosen to act as executive agents for the council in reviewing certain aspects of the RAND research effort. On occasion, certain council members have acted as super-project leaders for a cluster of RAND projects in a given area (e.g., J. F. Digby's role in RAND air-defense research and L. B. Rumph's work in strategic-offense studies). The RAND experience here thus seems to confirm in the microcosm what has recently been observed in the macrocosm, namely, that widespread "agreement on basic values" is not the indispensable cement to a stable political community that theorists have long fancied. Small group or organizational life may fluorish despite the absence of commonly shared purposes or objectives among the group or organization members. On the larger question of consensus in the community, see V. O. Key, Jr., *Public Opinion and American Democracy* (Knopf, New York, 1961).

sons who will provide representation for each of three classifications: (a) Industrial Trustees, who are from or representative of industry; (b) Academic and Scientific Trustees, who are from academic institutions or engaged in scientific research and development; and (c) Public Interest Trustees, who have an awareness and appreciation of the public and governmental interest in the corporation's operations. There has never been any direct government representation on the RAND Board, although at one time some thought was given to the possibility of reserving one or more of the Public Interest Trustee positions to government officials. The idea was rejected as being likely to lead to possible conflict-of-interest situations.[27]

The Board of Trustees meets semiannually, generally once in Santa Monica and once in New York or Washington, D.C. There are typically three days of meetings. Various aspects of the corporation's operations are discussed, and briefings on current RAND research projects are always given the trustees. During the period between meetings, a standing Executive Committee composed of the Chairman of the Board, the President and one other management trustee, and two nonmanagement trustees maintains an interim review over the corporation's operations and serves as a liaison body with the remaining trustees. In addition, a standing Finance Committee composed of three members of the Board of Trustees is empowered by the RAND by-laws to make investments or sell securities in the name of the corporation.

Article IV, section 11 of RAND's by-laws states that the Board of Trustees has full authority over all corporate activ-

[27] On several occasions, RAND trustees, when appointed to government positions, have left the RAND Board. Philip M. Morse, when appointed Director of WSEG, resigned his membership on the RAND Board; and William Webster did likewise when he became Chairman of the Research and Development Board.

183

ities. In practice, of course, much of the actual corporate business must be carried on by the full-time management officers acting under the Board's authority. Yet the evidence suggests that the RAND Board has a fairly active role in the broad direction of the corporation's activities, more so perhaps than most foundation trustees.[28] Whereas typical meetings of foundation boards seem to involve lengthy agendas dealing with financial transactions and other business matters, RAND's Board occasionally concerns itself with such important substantive issues as the orientation of the research program itself. On rare occasions, Board meetings have even served as a kind of review mechanism for evaluating the worth of particular studies. It is not unknown in the RAND experience for specific studies to have been effectively quashed by sharp questioning from knowledgeable trustees. Another important power of the Board, exercised through its Executive Committee, is the review of salaries paid by the corporation in excess of $25,000; all salaries in this range require the approval of the Executive Committee.

RAND trustees generally serve without pay, except that the Chairman of the Board receives an honorarium of $6,000 per year (the same as that received by the Chairman of the Ford Foundation Board of Trustees), Executive Committee members receive $2,500 per year, and other trustees receive travel expenses to board meetings and an honorarium of $200 per day for each meeting. Management members of the Board of Trustees receive no fees or honoraria by reason of being board members.

Elsewhere we have noted that the high caliber of RAND's trustees has been extremely helpful to the organization. Here it will suffice only to repeat that RAND has been fortunate

28 These observations are based largely on interviews with a number of RAND trustees.

in having men of distinction on its Board. It is to this factor perhaps more than any other that one can attribute the moderately influential role of the trustees in RAND's operations. Of particular note was the active leadership of H. Rowan Gaither who served (with one year's interruption) from 1948 until his death in 1961 as Chairman of the RAND Board of Trustees.[29]

TAX EXEMPTION PRIVILEGES

In general, the tax privileges enjoyed for a not-for-profit organization are of two principal sorts: (1) exemption from taxation of all income except that derived from unrelated business activities and (2) gifts made to tax-exempt organizations are deductible contributions.[30] Privileges of the latter

[29] In 1965 the RAND Board consisted of: Frank Stanton, President, Columbia Broadcasting System, Inc., Chairman; Frederick L. Anderson, Partner, Draper, Gaither & Anderson; F. R. Collbohm, President, The RAND Corporation; Marion Boyer, Executive Vice President of Standard Oil of New Jersey; T. Keith Glennan, President, Case Institute of Technology; J. R. Goldstein, Vice President, The RAND Corporation; Kermit Gordon, Vice President, The Brookings Institution; L. J. Henderson, Jr., Vice President, The RAND Corporation; William R. Hewlett, Executive Vice President, Hewlett-Packard Company; Edwin M. McMillan, Director, Lawrence Radiation Laboratory, University of California; Newton N. Minow, Partner, Leibman, Williams, Bennett, Baird, and Minow; Philip E. Mosely, Director, European Institute, Columbia University; Lauris Norstad, President, Owens-Corning Fiberglas Corporation; James A. Perkins, President, Cornell University; Kenneth S. Pitzer, President, Rice University; Don K. Price, Dean, Graduate School of Public Administration, Harvard University; David A. Shepard, Executive Vice President, Standard Oil Company (New Jersey); Charles Allen Thomas, Chairman of the Financial and Technical Committees, Monsanto Company; Charles H. Townes, Provost, Massachusetts Institute of Technology; and William Webster, President, New England Electric System.

[30] See F. Emerson Andrews, *Philanthropic Foundations* (Russell Sage Foundation, 1956), pp. 57-62, 92-125; Eleanor K. Taylor, *Public Accountability of Foundations and Charitable Trusts* (Russell Sage Foundation, New York, 1953), ch. iv; and Harry Graham Salter, *Fraud under Federal Tax Law*, 2 ed. (Commerce Clearing House, Inc., Chicago, 1953). In RAND's case, examples of exempt income are fee income from government contracts and income from investments as well as other incidental income such as book royalties and the sale of computer time.

sort have no applicability to RAND's situation, since RAND has never received gifts from profit-making enterprises or other donors which would otherwise be taxable income. For an organization like the Hudson Institute, however, this tax consideration is of considerable importance. (The Hudson Institute is a nonprofit research institution, patterned after RAND, established in 1961 by former RAND physicist Herman Kahn.) For Kahn intends to finance the Hudson Institute in part out of charitable contributions from industry.[31]

Tax privileges of the former sort, however, apply in RAND's case and are of substantial importance to the organization. To illustrate this point, let us consider a typical RAND fiscal year and see how the exemption privileges actually benefit the corporation. For the RAND fiscal year ending 31 December 1960, RAND received $875,343.07 in fees from U.S. government contracts, earned $68,639.34 in interest on corporation investments, and obtained $20,158.49 in royalties from RAND publications. Adding a few other miscellaneous sources of income and deducting certain nonreimbursable operating expenses, a net income of approximately $1,000,000 remained.[32] RAND expended part of this during the year on RAND-sponsored research projects, and retained part for working capital purposes. Now had this income been "profits,"[33] it would have been taxed at the normal rate of 30 percent for corporate profit under $30,000 plus a surtax of an additional 22 percent for corporate profit

[31] Interview, Herman Kahn, Santa Monica, California, August 24, 1962, and November 1962, Harmon-On-Hudson, New York.

[32] Figures obtained from internal RAND documents prepared for RAND trustees, September 1, 1961.

[33] Federal tax laws define "profits" as revenue minus expenditures for all business corporations organized to make financial gain and distribute dividends to shareholders.

in excess of that amount. As is readily apparent, tax exemption provides real benefits to the organization and loss of exempt status could have a serious effect on RAND's operations.

RAND is also exempt from state corporate income taxes by virtue of its incorporation as a nonprofit corporation under the General Nonprofit Corporation Law of California.[34] Yet, curiously, RAND does pay state and county property taxes on its real property. This fact illustrates the difficulty of making any broad generalizations about state and local taxation of nonprofit organizations. There is an enormous variety in the tax laws and court decisions of the several states. Many states exempt the real property of nonprofit organizations from state and local taxation. Some states do not. A number of states require property tax liability for property used in commercial enterprises even though the income is destined for use in an exempt organization's charitable activities. Other states exempt all property, however used, so long as the net income goes for the exclusive support of an exempt organization and its operations.[35] The same issue arises with respect to state corporate income-tax liability: are the "profits" of a business corporation liable for taxation if they are used exclusively for the exempt purposes of a tax-exempt organization? Under California law, a nonprofit organization may carry on a profit-making enterprise and continue to be a nonprofit and exempt organization provided that all income is channeled to the support of the exempt purposes. Other states, however, have different provisions in their tax laws.

Federal tax law presents a similar issue; and it is doubtless a far more important one since federal taxation repre-

---

[34] Title 1, Division 2, Part 1 of the California Corporations Code.
[35] Andrews, *Philanthropic Foundations*, pp. 61-82.

sents a far heavier tax liability. To what extent can an exempt organization engage in profit-making activities without (a) being subject to taxation on those activities or (b) running the risk of losing its federal tax exemption altogether? This issue, in fact, constitutes one of the several ambiguities surrounding the tax-exempt status of a nonprofit corporation like RAND (as well as many other nonprofit organizations, such as colleges and universities).

In the face of rising costs, colleges and universities have become involved in a variety of investment and business practices largely unknown several decades ago. Foundations established for New York University, for example, purchased four business corporations with the objective of channeling earnings to the support of the university. A consequent estimated loss in federal revenue resulted of approximately $1.5 million annually.[36] This exemption was upheld in a significant court case on the grounds that charitable destination was more important than source in determining an organization's tax-exempt status.[37] However, the law was changed shortly thereafter in the Internal Revenue Act of 1950. A chief objective of this act was to tighten up on the provisions concerning tax-exempt organizations in the internal revenue code. No longer could a tax-exempt organization carry on or sponsor profit-making ventures on a tax-free basis unless such activity was "substantially related" to the tax-exempt purpose of the organization.[38] The language of Section 101(6) of the 1950 Revenue Act, with a few minor amendments, remains

[36] Taylor, *Public Accountability of Foundations and Charitable Trusts*, p. 81.

[37] *C. F. Mueller Co. v. Commissioner*, 190 F. 2d 120, 1950.

[38] There were, however, some exceptions written into the law. Dividends from investments, for example, were excluded from the application of this provision and still qualified as exempt income. See Andrews, *Philanthropic Foundations*, pp. 399-403, for a list of the exceptions and the wording of the relevant portions of the revenue code.

the basic specification of organizations that qualify for exemption:

Corporations, and any community chest, fund, or foundation, organized and operated exclusively for religious, charitable, scientific, literary, or educational purposes, or for the prevention of cruelty to children or animals, no part of the net earnings of which inures to the benefit of any private shareholder or individual, and no substantial part of the activities of which is carrying on propaganda, or otherwise attempting, to influence legislation.[39]

In addition, the act specified that tax-exempt status could also be lost if an organization "unreasonably accumulated"[40] income or engaged in certain prohibited transactions (for example, paying any compensation in excess of a "reasonable allowance" for salaries or other personal services rendered). Despite this effort at clarification, uncertainties in the law have persisted. The law itself is rather ambiguously worded, and leaves few explicit guidelines for the courts or for administrators in the Bureau of Internal Revenue. Such phrases as "substantially related," "unreasonable accumulations," and "organized and operated exclusively," in the words of Eleanor K. Taylor, "have conjured up legal and accounting nightmares" which the courts and administrative rule-makers have been called upon to interpret.[41]

[39] In the 1954 Revision of the revenue code "testing for public safety" was added as an exempt purpose, and the proviso was also added that an exempt organization shall not "intervene in (including the publishing or distributing of statements) any political campaign on behalf of any candidate for public office." The section became Section 501(c)(3) of the Internal Revenue Code. See Andrews, *Philanthropic Foundations*, Appendix B.

[40] Exemption would be denied to organizations of the foundation type, whether trust or corporation, where amounts accumulated out of income in the current and prior years "are unreasonable in amount or duration in order to carry out the charitable, educational, or other purpose or function constituting the basis for exemption." Formerly Section 3814; in rhw 1954 Internal Revenue Code, Section 504(a)(1). Andrews, *Philanthropic Foundations*, p. 398.

[41] Taylor, *Public Accountability of Foundations and Charitable Trusts*, p. 86.

An illustration of the problem of judging when income is "unrelated" to exempt purposes and therefore subject to taxation may be seen in RAND's experience with a trailer court that occupied portions of RAND land for a time. When RAND purchased certain lands in 1958 for the eventual construction of new buildings, management decided to continue a lease arrangement with the trailer court that had occupied the land for several years prior to RAND's purchase. There were several principal motivations behind the decision. RAND officials did not want to incur the wrath of Santa Monica's elderly citizens by evicting a number of them on short notice (the trailer court was occupied mainly by retired couples). Moreover, RAND officials determined that perhaps twenty thousand dollars in annual rental income could be obtained by continuing the lease arrangement until construction was ready to begin. An arrangement was thus entered into with the trailer court for continued use of the land. After the agreement had been in effect approximately a year, however, RAND attorneys recommended against its continuance. They argued that such income might be construed as unrelated business income, therefore subject to taxation, and possibly liable to cause trouble with RAND's status as an exempt nonprofit organization. RAND management thereupon decided to terminate the trailer court's lease.

In contrast, an income-producing activity that is clearly related to RAND's exempt purposes would be royalties from the publication of RAND books. Since producing research work for publication is a part of RAND's fundamental aims, any income derived incidentally from the publication of books would presumably remain immune from taxation. Sale of computer time falls in this category as well. Since 1962, however, RAND has no longer claimed moneys derived from the sale of computer time as RAND income. At that time, in a gesture of good will during a critical juncture in contract negotiations, RAND voluntarily renounced such income. Henceforth, all funds derived from the sale of computer time were to accrue to the Air Force.

Another uncertainty in RAND's tax situation concerns the status of possible future income derivable from royalties on RAND patents. Since RAND's inception, it has had 200 invention disclosures, twelve issued patents, and at present fifteen pending patent applications.[42] At some future date RAND may possibly begin to receive income from royalties on these patents, and the question arises as to whether such income is "substantially related" (like book royalties) to RAND's main purposes or whether it is "unrelated business income" and thereby subject to taxation. RAND lawyers are currently grappling with this issue. Like other unsettled questions of this nature, this issue will likely remain to be worked out when the occasion demands in negotiations with the appropriate Bureau of Internal Revenue authorities.

A third uncertainty in RAND's tax situation warrants brief mention. This issue has not yet posed any actual problems for RAND, nor does it appear likely to become a serious issue in the near future. Yet it is the kind of latent background concern that an organization like RAND must be careful not to activate into a real problem. I refer here to the prohibition in the revenue code against "unreasonable accumulations" on the part of exempt organizations. Since the end of the 1950's, RAND has shown a tendency to build up its capital reserves to help cushion the shock of potential sudden fluctuations in the level of sponsor support. Such accumulations must be watched and not be allowed to grow to a point where

---

[42] Internal RAND paper. I am indebted to John Hogan, RAND legal historian and Director of Patents and Book Publications, for discussing with me the legal and tax status of RAND books and patents.

they appear to divert funds away from RAND's main charitable purpose: to carry on research in the public interest. What constitutes an "unreasonable" accumulation remains rather unclear, however, and in RAND's case there are few precedents for deciding when an accumulation is or is not unreasonable.[43]

In similar fashion, the statutory language against "carrying on propaganda, or otherwise attempting, to influence legislation" constitutes a potential question mark in RAND's tax-exempt status, and provides government tax authorities with a somewhat ill-defined potential element of control over RAND's activities. Is carrying on policy-oriented research "attempting to influence legislation?" Would it become so if, say, a group of RAND researchers sought to persuade congressmen that a certain policy course would advance U.S. interests?[44] No difficulties have arisen under this provision of the revenue code for RAND yet, and it is probably very unlikely that any will arise. Carrying on background research seems to be only in the most peripheral and indirect sense an "attempt to influence legislation."

A research organization like the Stanford Research Institute (SRI) has an even more intricate tax situation than RAND. Compared to SRI, in fact, RAND's situation is almost idyllic. SRI's difficulty stems from the fact that it engages in research not only for government clients but also for a variety of industrial and other profit-making concerns. The latter type of research to a certain extent is of a proprie-

[43] For a discussion of some standards that may help determine reasonableness, see Andrews, *Philanthropic Foundations*, pp. 95-98.

[44] RAND researchers have on occasion been wary of close contacts with Congressmen for this reason. In the legislative debate on the communications satellite bill in 1962, for instance, RAND staff members who had done research in the area maintained a cautious distance (but did offer opinions on the merits of the proposed legislation to a number of congressmen and committee staff members when asked for their views).

tary nature. That is, information is generated for the use of a specific industrial client and is not freely available to the general public.[45]

Several years ago, SRI volunteered to review its tax-exempt status in the interests of clarifying the ambiguity in the tax status of an organization that does both research for the government and proprietary research for private firms. Agreements subsequently worked out with Internal Revenue authorities and the Senate Finance Committee led to SRI's segregating out income derived from its commercial-type proprietary research operations from its other sources of income and agreeing to pay taxes on the former. In 1961, SRI began paying taxes on such income.[46] To segregate out income derived from and costs incurred in one aspect of the research program from the income and costs of the total research program constitutes a formidable accounting and administrative burden for an organization. SRI deserves credit for taking the initiative and cooperating fully with the Internal

[45] Section 512(b) (9) of the Internal Revenue Code states that: ". . . in the case of an organization operated primarily for purposes of carrying fundamental research the results of which are freely available to the general public, there shall be excluded [from the application of the provisions concerning unrelated taxable business income] all income derived from research performed for any person, and all deductions directly connected with such income." Presumably, income derived from research which is not "fundamental research" and the results of which are not "freely available to the general public" would not be excluded from taxation.

[46] Information obtained from interviews and documents at the Stanford Research Institute, Menlo Park, California, July 11, 1962. The tax paid the first year was small—$40,000. For a discussion of this development, see "The Profitable 'Non-profits,'" *Chemical Week*, May 6, 1961, p. 5. The editors of the journal commend SRI for its action, but criticize the agreement as not going far enough. They suggest that the distinction between "proprietary" and "public interest" research is not the best criterion to distinguish taxable and nontaxable income of the nonprofit research institution. Rather, they urge, the distinction should be drawn on the basis of the client alone: all income derived from research for the government should be exempt and all income derived from research for nongovernment clients should be taxable.

Revenue authorities to clarify an uncertain point in the nonprofit corporation's tax status. This action may well set a precedent for other nonprofit institutions engaged in both proprietary industrial research and contract research for the government.

But the important general point to draw from the present discussion is this: the tax-exempt status of the nonprofit corporation is hardly an automatic and irrevocable right enjoyed by any group wishing to proclaim itself a "nonprofit" organization. Tax-exempt status can be lost if an organization fails to observe defining conditions laid down by law, administrative action, and court decision. The withholding or withdrawal of tax-exempt status gives the federal government a powerful tool to regulate the activities of nonprofit corporations.

► VI ◄

RAND IN OPERATION: A CASE STUDY

The objective of this chapter is to contribute to a broad understanding of the advisor's role in policy formation through a case study of the origins, execution, and eventual communication to Air Force policy makers of one of RAND's most significant studies: the Strategic Bases Study (RAND Report R-266).[1] An incidental advantage of selecting the basing study (R-266) for analysis is the fact that most of the relevant materials are now free from security restrictions and can be discussed in the open literature. More important, the study offers an excellent illustration of a useful RAND systems analysis and how it can influence public policy. Tracing the study's evolution also throws important light on the nature of RAND and how it has operated.

A particular virtue of the case approach used here is the degree of realism and specificity it can give to our analysis of the policy-making process and the advisory group's role in that process. Generalized discussions of decision making with little empirical content often have an air of unreality about them. It is perhaps due in part to the absence of detailed

---

[1] A. J. Wohlstetter, F. S. Hoffman, R. J. Lutz, and H. S. Rowen, *Selection and Use of Strategic Air Bases*, The RAND Corporation, R-266 (April 1954), declassified 1962. A version of this chapter has previously appeared as "Strategic Expertise and National Security Policy: A Case Study," *Public Policy*, XIII (Yearbook of the Graduate School of Public Administration of Harvard University; 1964), pp. 69-106.

studies specifying the advisor's role in actual decisions that some misleading and even fanciful notions have arisen concerning the relationship between the scientific advisor and the policy maker. The view presented here sees the advisor as playing a vital, yet subordinate, part in the policy-formation process. In a complex technical age, the task of the defense policy maker can be enormously facilitated by analytic advice. Yet no amount of research or advice can absolve the policy maker from making a painful choice at some point: he must synthesize various technical and analytical considerations into a policy context which also includes value judgments, imprecise knowledge gained through experience, and intuitive estimates about such elusive entities as "human nature" or the "intent" of another nation. Rare is the research or piece of advice, especially at the broad strategic level, that can be translated directly into policy in some simple and straightforward fashion. Useful policy advice increasingly has become an effort to clarify choice under conditions of great uncertainty.

As a corollary to the above, it should be stressed that the notion of "scientific advisor" used here suggests much more than the traditional physical or natural scientists giving technical advice to policy makers on the performance of some missile, bomb, or other item of military hardware. What is distinctive about systems analysis is the interaction of various professional skills focused on a broad problem to achieve results beyond those of any single discipline. Thus systems analysis has contributed to the emergence of a new class of "generalist experts" who serve as something of a bridge between the worlds of science and technology and the world of policy. Significantly, the individuals who have played leading roles as generalist experts have frequently not been scientists or technologists. Social scientists of various kinds often seem to possess the type of integrating and synthesizing mind that equips them well for the task of bringing different specialized knowledge to bear on a broad problem. The traditional scientists may continue to have a major role in the making of national policy *for* science, for example, in the development of policies for the management and support of the national scientific enterprise, the strengthening of the country's basic science potential, and the selection and evaluation of substantive scientific programs. But on the issues of science *in* policy, basically political or administrative issues but influenced strongly by technical considerations, scientists have had to share their advisory role.

There are several reasons why it is particularly important to recognize the role of the economists and other social scientists. For one thing, this awareness serves as a healthy corrective to the view, popularized by Sir Charles Snow and others, that scientists in the strict sense (meaning physical and natural scientists and engineers) are possessed of some special gift of foresight in dealing with the "cardinal choices" facing the advanced industrial societies today.[2] Such views in their extreme form lead to a search for certainty—for intellectually pure solutions to enormously complex problems which admit of no certainty and no precise solutions deduced from a few simple initial premises. Further, they tend to undervalue the contribution of behavioral research and analysis to enlightened policy choices in the nuclear age. The cardinal choices of today are simply too complex, involve too many considerations beyond the competence of one trained in a particular scientific discipline,

2 C. P. Snow, *Science and Government* (Harvard University Press, Cambridge, Mass., 1961). For an effective critique of Snow, see Albert Wohlstetter, "Strategy and the Natural Scientists," in Robert Gilpin and Christopher Wright, eds., *Scientists and National Policy-Making* (Columbia University Press, New York, 1964), pp. 174-239.

and are too beset with large uncertainties to be resolved by an almost preternatural insight possessed by an elect of scientists and engineers. Tentative and empirical methods, reflecting the participation of numerous professional skills, are apt to be a more fruitful approach to the muddy normative issues of broad policy than the traditional scientific quest for objective solutions to clearly defined problems.

What the basing study shows above all is that it is possible to do useful empirical research on the cardinal choices which can be of great assistance to defense decision makers. The results of this work are not merely trivial; they are more reliable guides to decision than simple intuition, and they are vastly superior to either the civilian or military conventional wisdom.

A related reason why it is important to recognize the role of the social scientists concerns how one views the relationship between advisor and policy maker. Led astray by the assumption that the bases for making cardinal choices are accessible only to a chosen few, some observers could naturally conclude that the policy maker is almost helplessly dependent upon his scientific advisors. A very different view of the advisor's role is suggested, however, if the community of scientific advisors is understood to include a broader base of professional skills than simply the physical and natural sciences. Many more persons then have something useful to say about the cardinal choices, and they can say it in a language likely to be more intelligible and comprehensible to the policy maker. This is one of the salient differences between our present cold war situation and the circumstances in which decisions had to be made in World War II—a difference which the basing-study discussion brings clearly to light. The policy maker, civilian and military, is in a better position today to understand and to evaluate the advice he receives from his scientific advisors. His ability to make sophisticated decisions is enhanced by the filtering layer of generalist experts who help him to define the relevant policy alternatives springing from scientific and technological developments.

Moreover, he frequently has the time to acquaint himself with the data and the kind of reasoning required to give him a "first-hand knowledge of what those choices depend upon or what their result might be."[3] In a wartime situation, we would probably have a good deal more reason to fear that cardinal decisions would be made hastily and on the basis of unusual deference to scientists and other experts. But in present circumstances there is ample opportunity for debate, study, discussion, and review of advisory recommendations at a number of different policy levels. This has an important bearing on the question of how the scientific advisor figures in the policy process, and invites revision of elitist notions that the cardinal choices are steadily slipping out of the hands of accountable officials. The real problems in this area involve subtle questions of organizational pluralism. A major challenge will be to organize our advisory apparatus in such a way as to assure policy makers at various levels a broad base of scientific advice—while at the same time avoiding an extreme fragmentation in our decision-making machinery.

ORIGINS AND EARLY HISTORY OF THE STRATEGIC BASES STUDY

The origins of R-266 go back to May 1951, when the Air Force addressed a request to RAND for a study of the selection of overseas air bases.[4] Military construction authorized

---

[3] C. P. Snow, *Science and Government*, p. 1. For a fuller discussion of the importance of the time factor, see Wohlstetter, "Strategy and the Natural Scientists," pp. 181-183.

[4] The request came from Col. L. C. Coddington, Deputy to General Maddux, Assistant for Air Bases, Air Staff. The request for the study included a tenta-

by Congress for the fiscal year 1952 included some $3\frac{1}{2}$ billion dollars for air-base construction, almost half of which was for overseas base construction, and the prospect was that a much larger volume of new construction would be planned in the next several years. It appeared to the Air Force officer responsible for the request that RAND might make a useful contribution by studying the most effective ways for acquiring, constructing, and using air-base facilities in foreign countries. The criterion then in use for guiding decision on basing questions was a very crude one, having to do principally with minimum cost for given facilities. There was no concern shown for total systems costs which, as it turned out, were markedly different under alternative basing policies. The request was referred to Charles J. Hitch, head of RAND's Economics Division.

In keeping with the general RAND practice, Hitch sought to interest some of his staff in researching the area rather than attempting to thrust the project on anyone. The interest or lack of interest shown in a proposed study by the research staff would be an important factor in determining whether or not RAND would accept the Air Force's request. One of the men approached by Hitch was Albert Wohlstetter, a consultant of diverse background and interests newly added to the RAND Economics Division. Wohlstetter was not very interested in working on the project. "It did not look to me at the time," Wohlstetter recalls, "like a very interesting or challenging study . . . dull, full of nuts-and-bolts, the kind of thing one normally associates with logistics."[5] For a time, it

tive formulation of the problem and posed a number of questions for analysis in a short supporting staff paper.

[5] All direct quotations in this chapter that are not otherwise documented are based on personal interviews. My special thanks are due to Albert Wohlstetter, formerly of the RAND Corporation and now of the University of Chicago, for giving generously of his time and cooperating fully in this

appeared that the request for the base study would be one of those Air Force requests that RAND turned down. However, before giving a definite "no," Wohlstetter asked for a week or two to think about the matter. The week's reflection brought Wohlstetter to the conclusion that some potentially major problems might be raised by such a study. He opted to work on the study, and RAND informed the Air Force that it was willing to undertake the project.

Throughout the spring and summer of 1951, Wohlstetter was the only one formally working on the project. At this stage, Wohlstetter spent most of his time trying to formulate what exactly was the problem. He asked questions of himself and others constantly, and spent long hours familiarizing himself with Air Force procedures and current basing policies. He became convinced that the real task lay in discovering what were the right questions to ask instead of accepting the client's tentative formulation of the problem and providing answers to ready-made questions. Conversations with Henry S. Rowen, a RAND economist trained in engineering, were particularly helpful at this stage. Rowen was then engaged in several RAND projects related to Wohlstetter's field of interest, which made him a particularly valuable collaborator in the research. Later another economist, Frederic S. Hoffman, and an aeronautical engineer, Robert J. Lutz, joined Wohlstetter and Rowen.

While acknowledging Wohlstetter's and his team's primary responsibility for R-266, it should be stressed that they bene-

---

case study's preparation. Dr. Wohlstetter's personal letters and papers provided a useful source of documentation for many of the points discussed in this study. I also benefited from interviews with Henry S. Rowen, Frederic S. Hoffman, Robert Belzer, Frederick Sallagar (formerly special assistant to Secretary of the Air Force Thomas Finletter), and L. J. Henderson, Jr., of the RAND Corporation, and personal correspondence with Gen. Thomas D. White (USAF, ret.).

fited throughout from functioning in an environment that enabled them to draw on the skills and knowledge of others. The Wohlstetter team was able to solicit brief memoranda from RAND colleagues in the Electronics, Cost Analysis, and Mathematics areas on selected aspects of the basing problem and to interest some engineering people in making various calculations. Complementary studies in progress also provided a fund of data without which it would have been difficult to carry out the basing study.

INDIVIDUAL RESPONSIBILITY FOR STUDIES

The final report R-266, incidentally, bears the name of these four men as authors. The fact of individual authorship credit is not without significance, for it underscores an important point about RAND studies, namely, that they are the efforts of *individuals* and not the products of an abstract corporate personality. This means, among other things, that the corporation does not feel obliged to take an official RAND position on every study or piece of advice from individual staff members. When RAND issues a formal recommendation to a client[6]—or on occasion with a major study dealing with some particularly important or controversial problem—the corporation will take an official RAND position and the client is assured that the views of top management are represented.

[6] The formal recommendation is a special communication device that is used on rare occasions when RAND management wishes to bring some important matter to a client's attention. The formal recommendation consists of a letter from RAND's President, Franklin R. Collbohm, to a representative of the sponsoring agency recommending a specific course of action. The batting average for getting some sort of client action in response to a formal recommendation has been quite high. As of September 25, 1956, RAND had issued 49 formal recommendations to the Air Force and 36 had been accepted either fully or in part (source: unclassified portion of internal RAND materials, dated September 25, 1956). A formal RAND recommendation was not considered necessary in the basing study's case because the study itself contained numerous policy suggestions.

However, in most cases RAND takes responsibility only for assuring that the work represents a certain minimum standard of professional quality. Thus it is not uncommon for RAND studies or informal advice from different RAND people to conflict. This is not necessarily a cause for alarm. RAND's situation is not at all analogous to that of a decision-making agency responsible for a unified policy position. For a research organization, internal competition and dissension are healthy up to a point.[7] It is particularly desirable to avoid a "party line" to which individuals must conform. An "alive" intellectual atmosphere seems to require sustained, and sometimes acrimonious, debate about methods, standards of quality, and relevant problems for investigation. The organization like RAND does face some special problems in this respect because its status as institutionalized critic—with close ties to the "establishment"—means that nonintellectual standards of judgment will sometimes complicate internal debates about the quality of particular research projects or proposals.[8]

It is also important to note that the RAND practice of individual authorship credit contrasts sharply with the tradition of anonymity in government staff papers and reports.

[7] See Chapter V, note 3.

[8] This, of course, is not unique to the organization like RAND. Every research organization that lacks a completely secure financial base, or whose research product is oriented to a particular clientele, will reflect the impact of "market" pressures to some extent regardless of its institutional setting. For a discussion of some of the problems of a university research center dependent upon foundation financing, see Warren G. Bennis, "The Effect on Academic Goods of Their Market," *The American Journal of Sociology* 62:28-33 (July 1956). The problem here in large part is to maintain a high level of professionalism to be able to resist the potentially damaging effects of market or client pressures. Interestingly, RAND, with the relatively stable support of a major sponsor, seems in a better position institutionally to resist such pressures than the advisory organization which is supported by a large number of small contracts from numerous different clients.

This is not an inconsequential factor in considering how to create conditions for attracting and keeping research talent. A traditional feature of scholarly inquiry of all sorts is a strong desire on the author's part for recognition of his work.[9] When this opportunity is denied, as it is in many phases of government work, researchers may find the atmosphere uncongenial. They may prefer to work under conditions where they can receive recognition for a study, and are under fewer constraints as regards possible future publication of their work in scholarly journals. Thus the organization like RAND can serve a useful purpose in making a wider range of talent available for work on government problems than could normally be recruited for work directly within the government.

Yet RAND is not without its own problems in this regard. The fact that much of RAND's work is classified poses some difficult problems. For one thing, the "audience" with access to a classified report or study is typically much smaller than what one would reach in unclassified work. Since it is not always possible to prepare an unclassified version of a classified study for publication in the open literature, this can mean that on occasion an important study will be unknown outside a small community of cleared individuals. Second, a source of much irritation to people working on classified studies is the practice of outsiders' occasionally publishing in the open literature, and receiving credit for, ideas developed earlier by others in classified work. For this reason the academic strategist with access to classified materials is particularly resented by some strategists working primarily in the classified arena.[10] Finally, classified research

9 Robert K. Merton, "Priorities in Scientific Discovery: A Chapter in the Sociology of Science," in Bernard Barber and Walter Hirsch, *The Sociology of Science* (Free Press, Glencoe, 1962), pp. 447-485.

10 This is also a source of considerable tension within RAND. The re-

often tends to be more ephemeral than research published in the open literature, and is poorly indexed and difficult to gain access to, with the net result that a thoroughly professional tradition is lacking in classified research.

A DRAFT OF THE STUDY APPEARS: THE HARD JOB OF FILLING IN THE GAPS AND CHECKING THE ANALYSIS

By the late fall of 1951, Wohlstetter's thinking had crystallized to a point where he felt able to put some ideas down on paper. With Rowen, he drafted a "D"—or internal working paper—for internal RAND consumption which was completed on 29 December 1951—D-1114, "Economic and Strategic Considerations in Air Base Location: A Preliminary Review." This paper, though a document of some 100 pages in length and containing 40 pages of graphs and tables, represented only a sketchy summary of some useful approaches to the basing problem. But it did lay down in preliminary form what were to become the central concerns of R-266. Among the most important of these was the question of the vulnerability of aircraft on the ground to surprise atomic attack. It occurred to Wohlstetter that the past thinking on strategic bombing and basing posture had not given adequate attention to a question that could become of vital importance in the future: what would happen if the enemy struck first, hitting U.S. bombers on the ground before they reached enemy air space? The full importance of this question was only vaguely seen at this point, but certain uncomfortable conclusions emerged even from the preliminary analysis. It was found, for example, that Air Force regulations on base installation called for the concentration of

---

searcher who works in areas that enable him to publish in the open literature and gain an independent reputation is sometimes bitterly resented by colleagues who work mainly on classified projects. There is an informal feeling in RAND that everyone should do his "share" of classified work.

facilities to minimize the costs of utilities, pipelines, roads, and normal peacetime operational costs. Wohlstetter and Rowen discovered that such a system was highly vulnerable to enemy nuclear attack and suggested tentatively that a policy of dispersing facilities on bases would be preferable.

D-1114 contained many gaps and was inconclusive at points, but it had at least raised some significant questions and suggested new approaches to the problem of basing policy. For a variety of reasons, the study encountered considerable opposition and skepticism within RAND. Earlier RAND strategic bombing studies had dealt with such questions as: what is the best way to penetrate enemy fighter defenses? How high should bomber aircraft fly? What kind of aircraft, turboprop or turbojet, should be used? The question of where the strategic strike force should be based was not considered a very important issue. The offensive capability the Soviet Union had at the time was considered likely to be used only against cities, not bases, and this could be taken care of by appropriate air-defense measures. It was assumed that U.S. aircraft could be based at a variety of points within range of enemy targets at minimum risk. Because of this assumption many RAND people did not consider D-1114 to be dealing with a very significant problem. Some at RAND even believed the project was a waste of time and money, and should be discontinued. On the other hand, a few RAND staff members were impressed by D-1114, and thought the matter important enough to be briefed at the Pentagon without delay.

At this juncture, RAND's permissive and decentralized management policies played an important role in preventing either a premature cut-off of the project or a hasty effort to bring the findings to the Air Force without adequate verification. The decision was made at the divisional level by Economics Chief Hitch that the study raised enough interesting possibilities to warrant further investigation. Top RAND management served a useful function by not forcing a decision on the research staff at this early point. The decision of when to bring the results of the research to the Air Force's attention was left to Wohlstetter himself for the time being. Wohlstetter opposed any effort to communicate the preliminary findings to the Air Force on the grounds that further work was needed before any firm policy recommendations could be drawn from the research.

For the next several months, Wohlstetter and his small team (by now Rowen and Lutz were working more or less full-time on the study) spent long hours rechecking the assumptions of D-1114; calculating the effectiveness and costs of alternative basing postures under a variety of different conditions; interjecting new variables into the analysis and estimating their effect on the data; and determining which base systems would be most affected by errors in assessments of uncertainties (for example, if enemy capabilities were greater than anticipated by a factor of 10, or if mechanical failures were to cause a much higher rate of aborts than expected). Among other concerns, the study team gathered data on transportation and manpower costs at various base locations; assessed the costs of extending the radius of attacking aircraft by specified distances; obtained some judgment in conversations with State Department officials of the political problems involved in getting countries to furnish bases; and determined the accessibility of different base systems to alternate routes for penetrating the Soviet Union. The cost of achieving a given level of destruction of enemy targets under varying basing systems was calculated, as was the absolute amount of damage that could be inflicted on the enemy with different basing systems assuming a fixed budget available for the mission. The warning times available at each base in the event of enemy attack were also assessed, along with an analysis of the feasibility and cost of various means to extend warning time. The analysis generally proceeded by a method of successive approximations. Starting out with rather simplified assumptions, the analysis successively considered new constraints and problems of "real world" behavior likely to have an important influence on alternative basing systems. The RAND team throughout was concerned only with *gross differences* in system performance, that is, differences that would still be important despite large elements of uncertainty in the analysis.[11]

By the end of the spring of 1952, Wohlstetter and Rowen had completed a draft of the study. The 400-plus pages of the draft, and the numerous supporting papers used in its preparation, attest the extensive detailed and empirical work required to carry out a complex study. This time the results of the work appeared to be conclusive and to contain far-reaching policy implications. The analysis pointed toward the shattering conclusion that in the last half of the 1950's the Strategic Air Command, the world's most powerful striking force, faced the danger of obliteration from enemy surprise attack under the then-programmed strategic basing system.

Meanwhile, concern had gradually begun to grow in the Air Force over the implications of the Soviet A-bomb for the security of U.S. forces and problems such as air defense were receiving increased attention. A group around Secretary of the Air Force Thomas Finletter was particularly active in calling attention to the potential dangers of the Soviet Union's

---

[11] For an extended discussion of the basing study's methodology, see Albert Wohlstetter, "Analysis and Design of Conflict Systems," in E. S. Quade, ed., *Analysis for Military Decisions*, The RAND Corporation, R-387 (November 1964), pp. 103-148, and E. S. Quade, "The Selection and Use of Strategic Air Bases: A Case History," in *ibid.*, pp. 24-63.

acquisition of a nuclear capability and in urging that the United States begin planning for the time when the Soviet Union would have developed a substantial force of long-range bombers. In late 1951 Secretary Finletter wrote to Air Force Chief of Staff Hoyt S. Vandenberg suggesting that the Air Force undertake a major study of the vulnerability of SAC bases to enemy nuclear attack. In May 1952, a request came to RAND from General Craigie (which carried the imprimatur of the Air Force Chief of Staff) for RAND to study the problem of SAC-base vulnerability. Like the original request for a study of basing policies, the Air Force request only vaguely groped to define the issues. But as in the basing-study case the request showed the responsible official's genuine concern and sense for what is important, which often generates significant problems for detailed study. A special RAND study team on vulnerability was thereupon formed under the direction of mathematician H. Igor Ansoff. Although this team's work was eclipsed by the base study and no final report was ever published or issued to the Air Force, some of its work provided useful data for integration into the final version of the basing study. One of the researchers active in the other project also proved useful in the process of communicating the base study's findings to the Air Force.

It is also important to point out here the occurrence of a natural disaster in the fall of 1952 that indirectly had a bearing on the basing study's fate.[12] As often only disaster can, the event aroused high-level concern and dramatized the importance of a problem that was hitherto only dimly recognized. The event in question was a tornado that ripped

[12] The historical records at headquarters SAC report that the disaster occurred at Carswell Air Force Base on September 1, 1952. Personal correspondence, R. L. Belzer, The RAND Corporation.

through a major U.S. military airbase, completely destroying 12 B-36 heavy bombers. Since the bombers were thought to be secure, the disaster added to the worry over aircraft vulnerability already present in some Air Force circles and helped fashion a frame of mind receptive to the basing study's suggestions.

The last phase of work on the study marked the beginnings of the stage of communicating the research results to the client (and also the beginning of the sizable feedback of questions for further study). In talking to large numbers of Air Force officers during the middle and late months of 1952, Wohlstetter and his associates began injecting the ideas of the study into the Air Force hierarchy and through personal contact forming allies who were later to prove important in getting the results of the study incorporated into Air Force policy.[13] Before turning to that part of the story, however, let us take a close look at the study itself.

THE STRUCTURE AND CONTENT OF THE STUDY

As well as having broad implications for national-security policy, the strategic-bases study marked an important step in

---

[13] Indeed, R-266 owed much to Air Force officers in the various stages of groping to define important questions, design of the actual research and the strategy of communicating the results. Some of the Air Force officers who played particularly important roles were: General Maddux, Assistant for Air Bases; Colonel L. C. Coddington, Directorate of Operations (and, at the time of the study's initiation, Deputy to General Maddux); Lt. Colonel William M. Jones, Directorate of Plans and earlier of the Strategic Air Command; Colonel Stephen W. Henry, Directorate of Plans and Chairman of the Special Ad Hoc Committee to evaluate the RAND study; Colonel Maurice Stone, from the Assistant for Logistics Plans; Colonel William C. Moore, Directorate of Operations; and Major Charles V. Chapmen and Lt. Colonel Edward V. Munns, Directorate of Installations. It should be emphasized, however, that the basing study was not a collaboration with the military as was the SOFS project discussed in the previous chapter (or most of the WSEG studies where military personnel and civilian analysts work side by side on all phases of a project).

the evolution of the RAND systems analysis. It involved less elaborate mathematics than previous RAND systems analyses, which typically featured impressive methodologies and analytic techniques, combined with a large number of machine computations, aimed at finding the optimum solution to a specific problem. Instead, R-266 proceeded from the assumption that the elaborateness of the analytic techniques was not as important in policy-oriented analysis as the consideration, based on extensive use of empirical data, of the major factors and contingencies which affect choice on a broad problem. R-266 did not strive so much for an optimum solution —the best of all possible systems—as it did for a proximate approach that would identify a satisfactory system, capable of functioning well under widely divergent conditions and even giving some satisfactory performance under major catastrophe. While the RAND systems analysis became henceforth ostensibly less precise in method, at the same time it became more relevant to the policy maker's actual concerns. Further, in broadening the scope of the inquiry to include consideration of U.S. strategic objectives and such factors as political conditions of overseas base choice, R-266 figured prominently in the development of RAND's "strategic sense."

The objective of the study was to provide an analysis of how to select and use air bases for the strategic Air Force for the 1956-1961 period. Note that the time period with which the analysis was concerned extended nearly a decade into the future. The study involved considerations of the basing and use of weapons systems *which were not yet in operation* (and, in the case of the B-52, *not yet even in existence*).[14]

---

[14] The first prototype of the B-47 appeared in October 1951, and the B-47 was not included in SAC war plans until after 1953. Contrary to what is often supposed, however, the fact that a weapons system is not operational does not mean that there are no empirical data on which to base a system analysis

At the time the study was started, the U.S. strategic bombing force consisted of B-28's, B-36's, and B-50's. The focus of the study was to be the basing and deployment of the B-36, B-52, and especially the medium-range B-47; the latter was to be phased into the combat fleet beginning in 1953 and was destined to constitute the bulk of the strike force for the 1950's and sometime thereafter.[15]

The research strategy involved an exhaustive examination of four alternative basing systems in terms of their costs and their effectiveness in destroying enemy targets: (1) as a point of departure, the then-programmed system of bombers based in time of war on advanced overseas operating bases; (2) bombers based on intermediate overseas operating bases in wartime; (3) United States-based bombers operating intercontinentally with the aid of air-refueling; and (4) United States-based bombers operating intercontinentally with the help of ground-refueling at overseas staging areas. The examination of each of four alternative systems was carried out by taking certain principal factors which are important determinants of system cost and effectiveness and applying them to each system. These principal factors were the distances the bombers must fly from base to targets, to favorable entry points into enemy defenses, to the source of base supply, and to the points from which the enemy could attack these

---

of a future period. For a discussion of this point, see Albert Wohlstetter, "Strategy and Natural Scientists," in Gilpin and Wright, eds., *Scientists and National Policy-Making*, pp. 208-212.

15 The force was to be composed of approximately 1600 B-47's and RB-47's. Although R-266 recognized that basing policies had a bearing on R & D policy, it avoided any attempt to choose among the types of bombers available to perform the mission or already programmed for the force. This was done for several reasons. The study's major policy recommendations were essentially unaffected by differences in bomber technology. Also, the RAND team wished to disassociate the study from a bitter internecine Air Force quarrel between the proponents of a "big bomber" and those who favored the development of a smaller and faster bomber.

bases. The analysis was concerned with the joint effects of these respective factors on the costs of extending bomber radius; on how the enemy might deploy his defenses, and the number of our bombers lost to these defenses; on logistics costs; and on base vulnerability and our probable loss of bombers on the ground. The analytical treatment of the four critical base-location distances presented intricate problems and required skillful handling. On its face the problem appeared to involve contradictory elements. Considerations of logistics costs and ease of penetrating enemy defenses argue for locating bases close to the Soviet Union. Nearness to the source of supply and reduced vulnerability to enemy attack, on the other hand, argue for locating bases in or near the United States and away from the Soviet Union.

The results of the study showed that for a variety of assumptions the preferred system was alternative (4): United States operating bases for the strike force with the assistance of overseas refueling bases.

The then-programmed system of advanced overseas operating bases proved to be decidedly inferior to the U.S.-operated system with overseas bases used only for staging and refueling purposes.[16] R-266 demonstrated that this system would be extremely vulnerable to enemy attack in 1956 (even under very conservative estimates of Soviet forces). It would, in consequence, have the least destruction potential of enemy targets of any of the systems. Most of the projected overseas bases would be easily within the range of Soviet bombers.

---

[16] The then-programmed system envisaged the bulk of the U.S. strategic force being located in 30 bases on the continental United States in peacetime and then, upon the outbreak of war or a sharpening of international tension which presaged war, being moved overseas, to operate from a base system consisting of about 70 bases. Fighter defense and anti-aircraft battalions would be provided but there was relatively little emphasis given to the passive defense of this system.

Moreover, warning time of an enemy attack would not be sufficient to permit evacuation of our aircraft in time to escape destruction. Furthermore, even under favorable assumptions about the size of enemy stockpiles of atomic weapons and the yield of the weapons and the state of U.S. base-defense capabilities, it was found that, by a first-strike, the enemy could destroy almost the entire U.S. combat force while it was still on the ground. And the attack would make it unlikely that the small surviving part of the force could respond effectively and penetrate enemy defenses in a retaliatory strike. The cornerstone of U.S. policy at the time —deterrence of aggression through the nuclear striking power of the Strategic Air Command and destruction of enemy industrial targets if deterrence failed—was thus seen to be jeopardized by the projected basing system. Indeed, the whole concept of deterrence as it was then conceived seemed in need of revision. "Deterrence" was thought of largely as deterring a massive assault on Europe. R-266 showed that a vital part of any viable deterrence policy had to be deterrence of an attack on the deterrent forces themselves through the provision of a second-strike strategic capability. Thus the basing study contributed to important changes in U.S. strategic thought and doctrine. The insistence on a secure second-strike deterrent force later emerged as a central element in the McNamara strategic doctrine.

The second alternative of intercontinental operation with the aid of air-refueling (which had some strong advocates in the Air Force) was shown to buy lower base vulnerability at so high a cost that total striking power would be drastically reduced. Air-refueling, the study demonstrated, was much more expensive than conventional ground refueling for any one of a number of different possible air campaigns. Here one sees what is meant by *gross differences* in systems cost and performance, which was referred to above as one of the study's operational principles. The differences in cost between the air- and ground-refueling systems were not of a marginal nature—they were of the order of magnitude of 10-15 billion dollars over the life of the system.

The third alternative of intermediate overseas operating bases proved to be the worst of all the systems in that it combined some of the major disadvantages of the advanced-overseas-operating system and the air-refueling system. It would be almost as vulnerable as the former because even intermediate-distance operating bases would be within range of enemy attack and not within the warning network that permitted evacuation of aircraft in time to escape destruction. And, like the air-refueling system, it would be costly to operate because supplies and personnel would have to be moved overseas and expensive facilities at the bases constructed to accommodate the maintenance, bomb-loading and manifold other functions of an operating base.

The system relying on overseas bases for staging and refueling purposes was shown to be the "best" system on the grounds, first, that it was relatively invulnerable to enemy attack either before or after the strike against the enemy. U.S. bombers would only be on the ground a very short time for refueling either before and/or after striking their targets, making it very difficult for the enemy to destroy them on the ground. Routes could also be varied so that the enemy would never be sure which base the U.S. bomber force would use. Moreover, the bases would not need the large-scale construction of expensive and vulnerable facilities that would be necessary in an operating base. Modest expenditures on underground storage for fuel would radically reduce vulnerability to almost any atomic attack within the enemy's capabilities at that time. This conclusion was not sensitive to unfavorable resolution of some of the major uncertainties of the analysis. The results would hold true even if, for example, the estimates of the number of A-bombs in the enemy's stockpile were underestimated by a factor of 10 or if the bombs were of a larger yield than anticipated. Further, the total costs of the system were lower than the other three systems, thus freeing additional resources for the performance of the mission. Considered in a number of complex aspects, the overseas-staging-base system appeared to be markedly superior to the other three systems.

COMMUNICATING THE RESEARCH RESULTS TO THE FOCAL POINTS OF DECISION

In policy-oriented research the communication of results is almost as important a task as the research—and sometimes hardly less difficult and demanding. Paradoxically, the client agency may be strongly motivated not to use the research that it has sponsored. As Glock comments:

Offhand, it might be expected that the client, by virtue of his role, would function wholly to foster utilization. Having commissioned a research project, he, among all the parties concerned, would be the most highly motivated to use its results. Where utilization does not occur, therefore, one would be tempted to look elsewhere for an explanation. However, an examination of the record suggests, perhaps surprisingly, that the client is very often directly responsible for the nonutilization of the results of research which he sponsors.[17]

At the outset it is important to recall that decision making in the defense establishment (as elsewhere in the government) is a *process*. Phrases like the "decision-making process" and the "process of policy formation" are not mere

[17] Charles Y. Glock, "Applied Social Research: Some Conditions Affecting Its Utilization," *Case Studies in Bringing Behavioral Science Into Use*, Studies in the Utilization of Behavioral Science (Institute for Communications Research, Stanford University, 1961), I, 7.

216

incantation: they refer to the continuous flow of decisions, large and small, that make up the seamless web of policy formation and administrative action in the federal government. The dynamic flux of the process makes the job of the advisor particularly difficult. It means that there is no orderly procedure whereby the advisor can state his views or explain his research and then retire from the scene confident that his advice will receive systematic consideration. There are numerous distractions and competing demands on the decision maker's time and span of attention. Decisions once made can become unmade a week later. The advisor may face a difficult task to secure a full hearing for his views in the first place, and then must struggle to keep attention focused on his recommendations for a long enough period to assure action of some kind. *Continuity* is thus an essential attribute of effective communication of policy-oriented research.

A corollary of this is that the advice cannot simply be given to the top levels if favorable decision and effective implementation of the advice is desired. Consider the case of a high-level decision maker accepting the recommendation of an advisory group and making a "policy" decision designed to implement the advice. Unless the subordinates carry out the decision effectively, the whole intent can be defeated.[18] Comprehension of the basis for the decision

---

[18] "Goal displacement"—or deflection of an agency from its original or stated objectives to other ends—is a common theme sounded by analysts of bureaucratic behavior. Useful discussions of this theme include: Henry Wriston, "The Secretary and the Management of the Department," and Don K. Price, "The Secretary and Out Unwritten Constitution," in Don K. Price, ed., *The Secretary of State* (American Assembly, New York, 1961), pp. 76-112 and 166-190; Seymour M. Lipset, *Agrarian Socialism: A Cooperative Commonwealth Federation in Saskatchewan* (University of California Press, Berkeley and Los Angeles, 1959); Peter M. Blau, *Bureaucracy in Modern Society* (Random House, New York, 1956), esp. pp. 85-101; and Paul Appleby, *Policy and Administration* (University of Alabama Press, Birmingham, 1949).

reached at the higher level can be a vital factor in winning the consent and enthusiasm of those who must execute the decision and, in doing so, make a myriad of other decisions which can determine the success or failure of the original decision. It follows therefore that it is often desirable to communicate the research or advice to the working levels of an organization as well as to the higher policy levels.

In the case of R-266, the task of communicating the findings of the analysis to the Air Force lasted from about the fall of 1952 through most of 1953. By the end of 1952, the study had begun to assume final form. For some months, Wohlstetter and his colleagues had been in close contact with Air Force officers, checking out the assumptions of the study and beginning to circulate its potentially revolutionary conclusions. In January of 1953 the research had progressed to the point where Wohlstetter now felt confident that RAND should present the findings formally to the Air Force. Hitch at the division-head level agreed, and so did top RAND management. Pressure also began building up within the Air Force from sympathetic officers for an early release of the findings. The question then became one largely of tactics.

An initial concern felt by RAND management was the need for some special device or tactic to dramatize the study's importance so as to maximize its impact on Air Force policy. Accordingly, it was decided that the results of the study, by now nearly completed, would be disseminated to the client in the form of an unusual category of RAND publication—the staff report. This staff report (R-244-S) amounted to a condensed version which summarized the essential findings and policy recommendations of the larger study. The staff report offered the advantage of attracting more attention than one of the more frequently used publication categories, and its relatively short length seemed to assure a large readership. RAND management also decided that the staff report should be made a *special* staff report. This simply meant that for the time being the research results would be distributed solely to the Air Force. Here one gains some insight into what is meant by a confidential advisory relationship with the client. The Air Force had an initial opportunity to weigh the merits of the study on its own, and was protected against premature criticism from superiors in the defense hierarchy, a sister service, or other government agency while the study was under consideration.

With the distribution of 75 copies of R-244-S throughout the Air Force on 1 March 1953, the communication of the research findings entered into its most intensive phase. This phase lasted until approximately November of the same year.

Wohlstetter had made several trips to Washington in January and February 1953 in preparation for the report's release, but for the next eight months after the March 1 date of publication he was almost constantly in Washington. A large portion of that time was consumed in briefings. The briefing can often be a very effective technique of communication because it permits interaction between the researcher and the decision maker, and the researcher can answer objections to the study that the decision maker might raise. In the present case, the importance of the subject meant an extraordinarily large number of briefings and its startling conclusion foretold a vigorous interchange of views.

In fact, Wohlstetter gave 92 briefings (most of them during the period from March to the end of October). Some idea of the considerable interaction between briefer and audience is suggested by the fact that 16 charts were used for the main briefing, whereas it was found necessary to prepare 70 charts for use during the question period. The question period also typically lasted longer than the main briefing. Wholstetter was assisted in the grueling briefing routine by Hoffman, Rowen, or Lutz and sometimes by all three. In addition, RAND Vice President L. J. Henderson, Jr., RAND mathematician Robert Belzer (who had worked on the other RAND vulnerability project mentioned earlier), and RAND engineers J. J. O'Sullivan and J. C. DeHaven occasionally assisted in the briefings. Besides the group briefings, Wohlstetter was called into numerous confrontations and individual discussions of the study as a whole or various aspects of it. These facts give one some impression of the magnitude of the effort and the number of man hours involved to communicate the research results to the client. They also give one some idea of the earnestness of government officials who must take the responsibility for "living with" any decisions they make.

The initial briefing began with the Strategic Air Command (SAC), the functional command most directly affected by the study's recommendations. The immediate response was enthusiastic. The briefing (and report) came as a "shocker" and aroused great interest within SAC. Increased support was added to the initial group of sympathetic officers who shared Wohlstetter's conclusion that the vulnerability of SAC under the programmed base system was a serious threat to the nation's security. Surprisingly, very little opposition to the study was encountered at this point on the basis of its being done by civilians without "military experience."

Next Wohlstetter and his team went to the Pentagon to brief a group of about 40 senior colonels representing various directorates of the Air Staff. The reception was generally favorable. It was determined that a "saturation" campaign of briefings should be undertaken throughout various commands and major components of the main Air Force directorates with a view to testing the research findings and generating the momentum to bring them eventually to the Air Force

Council.[19] The saturation campaign of briefings continued through the end of May. Then in early June a significant briefing was arranged with officers of general rank. This was the highest ranking group that Wohlstetter had briefed thus far, and it was at this briefing that an important decision was taken. The generals decided to create a special Ad Hoc Committee of the Air Staff to check out the component parts of the study in terms of their accuracy, reasonableness of assumptions, and feasibility of implementation.[20] The intention was that the Ad Hoc Committee would examine the study in depth, and prepare a report for submission to the Air Force Council. If the Ad Hoc Committee submitted a favorable report, the prospects looked bright for favorable decision and action on the study's recommendations.

A number of factors, however, cautioned against any easy optimism. As happens frequently with review committees, the Ad Hoc Committee decided to operate on the basis of parceling out specific areas of interest to particular committee members. This meant that something of a system of "concurrent majorities" would obtain, with the Wohlstetter team having to persuade each committee member (and the directorate or functional command he represented) to endorse the

[19] The Air Force Council was attached in the Air Force hierarchy to the Vice Chief of Staff's office; it acted as a staff arm and decision-making aid to the Vice Chief of Staff. The Chief of Staff had the final authority to act or not to act on Air Force Council advice. Typically, however, a recommendation by the Air Force Council met with the Chief of Staff's approval and wherever possible the Chief of Staff encouraged consideration of important issues by the council. It might be noted, parenthetically, that it was at this briefing of senior colonels where Wohlstetter encountered his only serious objection to civilians proferring advice to military professionals. But several officers with extensive combat experience successfully rebutted their colleagues' objections to civilian assistance of this kind.

[20] The Ad Hoc Committee was also to absorb another Air Staff Ad Hoc Committee which had been organized separately in March 1953 to analyze the vulnerability question.

study. It was important to persuade each committee member (and each directorate) because separate "concurrences" were to be solicited from each directorate to form a major part of the Ad Hoc Committee's report. In effect, each member to some extent could exercise a check or veto on the committee's recommendations.

Meanwhile, the momentum that had developed toward a rapid decision on the study's recommendations had now begun to slow down. Jurisdictional questions arose, for example, which presented new obstacles to early decision on the study. SAC, as a specified command, was responsible to the Joint Chiefs of Staff for policy guidance. But at the same time it was still tied closely to the Air Force and had certain ill-defined responsibilities to the Air Staff. This divided responsibility between the Joint Chiefs of Staff and the Air Staff raised difficult questions as to the exact nature of SAC's relationship to the Air Staff. Elements within SAC began to fear that the study could be used as an opening wedge for the Air Staff to interfere with internal SAC operations and responsibilities.

Another troublesome problem was an internal Air Force quarrel between the "big bomber" proponents and those who favored the development of a smaller and faster bomber. At several points the basing study seemed in danger of being eclipsed by this quarrel or else reduced to a pawn in the struggles between the contending factions. The RAND team finally managed to avoid being submerged in the quarrel by demonstrating that the study's major policy recommendations were essentially unaffected by differences in bomber technology.

Elsewhere pockets of resistance developed as it was realized that acting on the study would involve drastic changes in programmed activity.[21] The inertia of established programs proved difficult to overcome even in the face of strong evidence arguing for innovation. A point of concern also was the prospect that substantial changes might be interpreted by rivals as an admission of error on a vast scale. The RAND team as well encountered opposition from a number of Air Force officers who genuinely feared that the drastic revision in policy suggested by the study could undermine the confidence and morale of their units. Conceivably the Air Force could also be embarrassed before Congress and might even become involved in a congressional investigation. Thus even some Air Force officers who agreed with the RAND study's recommendations felt it desirable to straighten out the error as unobtrusively as possible at some future date. In this general atmosphere, critics saw the special Ad Hoc Committee

[21] This is consistent with the March-Simon contention that organizational innovation will occur more easily if the proposed changes are of a gradual incremental nature—"when the carrot is just a *little* way ahead of the donkey —when aspirations exceed achievement by a small amount." If the proposed innovation is too drastic a departure from current doctrine or procedure, frustration results and "neurotic reactions interfere with effective innovation." James G. March and Herbert A. Simon, *Organizations* (John Wiley, New York, 1958), p. 184. To allay doubts, the Wohlstetter team on occasion sought to underplay the novelty of their recommendations.

The subject of organizational change has recently attracted increasing scholarly interest. The reader is referred to: March and Simon, *Organizations*, ch. vii; Warren G. Bennis, "A New Role for the Behavioral Sciences: Effecting Organizational Change," *Administrative Science Quarterly* 8:125-165 (September 1963); David L. Sills, *The Volunteers* (Free Press, Glencoe, 1957), ch. ix; Samuel P. Huntington, *The Common Defense* (Columbia University Press, New York, 1961), ch. v; Sheldon L. Messinger, "Organizational Transformation: A Case Study of a Declining Social Movement," *American Sociological Review* 20:3-10 (February 1955); James Q. Wilson, "Innovation in Organizations: Notes Toward a Theory," paper read at American Political Science Association, New York, Sept. 1963; William H. Starbuck, "Organizational Growth and Development," in James G. March, ed., *Handbook of Organizations* (Rand McNally, Chicago, 1965), pp. 451-533; and work in progress by Anthony Downs, The RAND Corporation and the Real Estate Research Corporation.

as a convenient vehicle for delaying action. Studying the study, it should be noted, often affords the decision maker a minimum-risk course of action: he can thereby avoid or post-pone action that might create enemies and at the same time appear to be doing something to satisfy critics. Hence those who, for one reason or another, did not favor adopting the study's recommendations now largely adopted the tactic of delay and attrition and opposition to any immediate action or decision on the study's main recommendations.

The delaying tactics had a notable success in July and August of 1953. Wohlstetter had the feeling that his "wheels were spinning" during this period. "It looked as though we would convince everybody intellectually," he recalls, "but that nothing would get done." Some of the sense of urgency had departed, and the RAND team found that something of a reverse "bandwagon effect" was beginning to take effect. A number of original Air Force supporters reversed themselves and either lined up with the opposition or adopted a non-committal position.

In the face of these setbacks, the RAND team and the Air Force group favoring the study intensified their efforts to communicate the research results. A planning exercise was even carried through which tested the recommended system's capacity to handle a complex strike. Despite redoubled efforts, the fortunes of the study continued in doubt. By the end of the summer, Wohlstetter was convinced that some drastic step was necessary to assure that the study reached the Air Force Council for consideration. Consequently, he proposed to RAND Vice President Henderson that a special visit be made to General Thomas S. White, acting Chief of Staff of the Air Force,[22] to focus high-level attention on the study and to guard

---

[22] General Hoyt S. Vandenberg, Chief of Staff, was ill with cancer.

against a permanent tie-up in delays and jockeying at lower levels. Henderson agreed, and an interview was arranged between General White and a delegation from RAND consisting of Wholstetter plus the top echelon of RAND management (President Collbohm, Vice President Henderson, and Vice President J. R. Goldstein). This interview proved to be an important turning point.

The RAND team was now assured of consideration of the study by the Air Force Council; and in any event they could now count on an appeal to the highest decision-making level in the Air Force. From this point on, the prospects for favorable decision brightened and the opponents of the study were placed on the defensive. Ironically, Wohlstetter and his colleagues also received a fortuitous assist from an unexpected quarter: Premier Malenkov's announcement that the Soviet Union had detonated a hydrogen bomb.[23] Although intelligence estimates at this time were uncertain, there was little disposition in the Air Force after the Malenkov announcement to doubt that the Soviet Union had a substantial nuclear capability. Wohlstetter and his colleagues capitalized on this announcement in the late summer briefings to dramatize the dangers of an enemy first-strike against vulnerable overseas bases.

Several aspects of the interview with General White deserve additional comment. First, White was influenced in his decision to place consideration of R-244-S on the Air Force Council agenda in part by the pressure of the budgetary cycle. For the advisory group interested in timing the release of reports and briefing campaigns to maximize their impact on policy, this is an important factor. Although the decision-

---

[23] The Malenkov announcement more than offset the negative effect of former President Truman's famous comment after he had left the White House that he doubted whether the Russians really had the A-bomb.

making process is diffuse and tends to be geared to crises which are by their nature unpredictable, there is always the assurance that the pressure of budget preparation will pose important policy choices and in general serve as a stimulus to decision.

Second, the fact that it was even possible to approach General White, when seemingly blocked or delayed at lower echelons, deserves notice. RAND's existence *outside* the Air Force hierarchy, and its reputation for independence and objectivity, is seen as playing an important role. It is doubtful whether an Air Force officer, an "in-house" advisory group made up of Air Force career personnel, or even a civilian advisory group attached to a unit within the normal chain of command, would have the same opportunity or incentive to by-pass immediate superiors and press for the adoption of controversial ideas at higher levels. The data here thus support the tentative hypothesis of Merton and others that the utilization of expert advice in the behavioral sciences will be positively influenced by location of the advisor outside the organization for whose procedures or policies he is suggesting innovation.[24]

It should also be stressed that location of the research or advisory unit outside the organizational framework of the decision-making agency allows greater scope for truly original and creative ideas to emerge in policy-oriented research. Within an agency responsible for policy making and operations, there are strong and understandable pressures to ensure agency-wide policy coordination and orientation of the work effort toward shared objectives. It is difficult for anything but

24 Robert K. Merton, "The Role of Applied Social Science in the Formation of Policy," *Philosophy of Science* 16:168 (July 1949) ; and Ronald Lippitt, "Two Case Studies in the Utilization of Behavioral Research," *Case Studies in Bringing Behavioral Science Into Use*, I, 34ff.

"symbolic" research, that is, research that is innocuous, does not reflect unfavorably on agency policies or procedures, and raises no disruptive problems to be conducted in such an organizational setting.[25] Thus it would appear both on grounds of accessibility to the various points of decision scattered throughout the client organization and on grounds of effective use of scarce creative-research talent that a strong case can be made for contracting many policy research and advisory functions to an outside organization.

This conclusion may have an important bearing on questions of government organization in the scientific age. As government agencies increasingly require various research and advisory services, experimentation with novel and decentralized organizational forms like the RAND Corporation (which are well-adapted to research activities) may extend to a growing number of nondefense agencies.[26] The traditional Weberian notions of the bureaucracy as a hierarchy of fixed offices may then seem antiquarian as a loose-jointed system of administration-by-contract becomes a prominent feature of government organization. The staff agency's function may come to resemble in many ways the management of a giant research establishment.

Last, it should be noted that there was no "end run" of the sponsor in this case. Clearly, problems of a fundamentally

[25] Joseph W. Eaton, "Symbolic and Substantive Evaluative Research," *Administrative Science Quarterly* 6 (March 1962); and Ashley L. Schiff, *Fire and Water: Scientific Heresy in the Forest Service* (Harvard University Press, Cambridge, Mass., 1962). Schiff, overstating his case considerably, even goes so far as to suggest at one point that research should be entirely divorced from an action agency. A useful commentary on the Schiff study is W. Eric Gustafson, "Science vs. Administrative Evangelism," *Public Administration Review* 22:84-88 (Spring 1962).

[26] See Chapter VIII below, and Don K. Price, "Creativity in the Public Service," *Public Policy*, IX (Yearbook of the Graduate School of Public Administration of Harvard University, 1960), pp. 1-17.

more complex nature would have been presented had RAND sought to by-pass the Air Force in the present case and bring the study results directly to the Secretary of Defense's attention. A striking feature of this case study, indeed, is the fact that strategic decisions of enormous importance were made solely by a military service. Now that the Secretary of Defense has assumed primary responsibility for strategic policy it is doubtful that such decisions could be made without involving OSD agencies and personnel.

A McNamara memorandum of 12 April 1962 has had a somewhat ill-defined, though potentially important, bearing on the question of "end runs" around a military-service client to a higher policy echelon. Secretary of Defense McNamara directed the Secretaries of the Air Force, Army, and Navy to forward to his office copies of "each report of study received by your Department from RAND Corporation, Operations Research Office, Operations Evaluation Group, Institute for Defense Analyses, similar research or analysis groups under contract with your Department . . . concurrently with receipt by your Department."[27] For a time, this McNamara order caused great alarm and shock throughout the military services and the contract-research organizations. The military services particularly were concerned that there might no longer be scope for private and confidential exchanges of information between the service and its advisory institution. The advisory institutions were concerned that there might be less of a chance to explore controversial problems for extended periods without arousing the attention of high-level policy makers. Time to think through the issues and to examine the feasibility of novel approaches, in an atmosphere relatively free from outside pressures, is an important requirement for

[27] Copy of the McNamara memorandum obtained from the Office of the Secretary of Defense.

subtle analytical studies. As we saw in the present case, Wohl-stetter and his colleagues were fortunate to have a relatively long "incubation period," out of the spotlight of official attention, in which their ideas could mature.

As of the end of the year, however, the April 12 memo-randum appeared not to have been applied strictly and not to have had the expected far-reaching consequences. Secretary McNamara seems to have been primarily interested in inform-ing himself and keeping abreast of the latest research on im-portant defense problems. He was apparently not interested in exercising direct control over the military services' policies toward the distribution of reports and studies from the service advisory institutions. Nevertheless, the problem of "end runs" around a service sponsor has become a more real and trouble-some one for the service-affiliated advisory institution. The researcher faces particularly difficult problems of profes-sional ethics on rare occasions when conflicts develop between loyalties to a service sponsor and loyalties to the larger national interest. And management generally faces a more difficult task in shielding the staff from client pressures and in assuring a sufficient "incubation period" for the genuinely provocative ideas to emerge.

THE IMPACT: R-266 REFLECTED IN AIR FORCE POLICY CHANGES

Distinguishing between the stage of communicating the re-search results and the stage of their actual reflection in Air Force policy is somewhat arbitrary. In a sense the study began to have an "impact" even before the RAND team dis-tributed copies of the special staff report and initiated the program of formal briefings. Numerous informal contacts existed between the RAND team and various parts of the Air Force, and the thinking of many Air Force officers had already been affected by contact with the ideas contained in the study. Indeed, in many cases the advisor's most important contribution is simply to force a rethinking of some complex problem. The end product of most planning and research activities is not an agenda of mechanical policy moves for every contingency—plainly an impossible task—but rather a more sophisticated map of reality carried in the minds of the policy makers. If the basing study had done nothing more than to demonstrate the shortcomings of the old base-selection policies, it would have made a very useful contribution. As matters turned out, however, the basing study was to do much more. The study was to have a direct and visible impact on Air Force policy to an extent that is unusual with analytical studies.

For present purposes, a "policy" change will mean concrete action, or decisions leading to concrete action, affecting Air Force plans, strategic objectives, operating procedures, installations practices, training requirements, or other tangible behavior to which the RAND research in some clear-cut fashion contributed. Attention will not be focused primarily on changes in attitude or outlook of Air Force policy makers (important though these might be).

There are a number of difficulties with assessing the policy impact of the advisor even given a restrictive definition of "policy." The fact that a policy change follows the submission of advice does not, of course, necessarily imply a causal relationship between the two. It is also evident that influential advice often may come from different sources. Witness, for example, the decision to accelerate plans for the development of the ICBM as a military weapon. A RAND scientist, Bruno Augenstein, and scientists of the Von Neumann Committee independently arrived at the conclusion that smaller weight warheads could make such a weapon system feasible. In such a case, it clearly becomes difficult to assign weights to the relative impact of the different advisors.

Added to these general considerations is the fact that the effective advisory group usually goes to great pains to conceal its impact on policy. It is not a sensible tactic to claim credit for particular decisions, since this will likely irritate, annoy, and even frighten those who must assume final responsibility for decisions. On the contrary, an important task of the advisor is to elicit the decision maker's involvement in the study's fate. Clearly, this can be accomplished more readily in most circumstances by making the decision maker feel the ideas are really reflections of his own thinking than by making him appear obtuse and intransigent. In the present case, the final publication of R-266 was held up to 1 April 1954, in part to permit the Air Force to implement some of the suggested changes on its own prior to the report's formal release. An opportunity was thus afforded for the client to have already initiated policy changes at the time R-266 was being circulated generally within the defense establishment. The final wording of the report also reflects this concern with the sensitivities of the decision makers. Phrases like "the *formerly* programmed" and "the *then* programmed system" were used to refer to the base system originally planned for the 1956-1961 period, and care was taken to note the efforts made by the Air Force to revise the original plans.

Nonetheless, despite the difficulties in any effort to identify the policy impact of a research product or item of advice, the evidence in the present case clearly suggests that the RAND base study was the catalyst to major Air Force policy changes.

A chain of personal interaction establishes beyond doubt that the ideas developed in the base study received the close attention of Air Force decision makers. Furthermore, an unmistakable relationship seems to exist in this case between the advice and the policy changes. The distinction between a first-strike and a second-strike capability—perhaps the main conceptual contribution of R-266 on which a number of its spe-

cific policy recommendations were based—seems to have
originated with the Wohlstetter team. Questions of intellectual priority are, of course, extremely difficult to assess.
Simultaneity of discovery has often marked the emergence of
new intellectual concepts.[28] But in 1953-1954 the distinction
between a first- and second-strike capability was essentially
a novel concept, although in various parts of the strategic
community increasing thought was being given to the possible
implications for U.S. policy of the acquisition of a substantial nuclear capability by the Soviet Union. At the very least
it can be said that the basing study (and follow-up research
by Wohlstetter and colleagues) was the first systematic expression of ideas vaguely present in the strategic community.
No other study or other group had so clearly recognized the
importance of a secure deterrent force and spelled out its
implications for U.S. policy. Several previous studies had
dealt generally with the problem of vulnerability, but none
had drawn explicit attention to the need for developing a
deterrent force capable of surviving an initial enemy atomic
assault and still inflicting unacceptable damage on the
enemy.[29] Nor had the operating procedures and strategic

[28] See Robert K. Merton, "Priorities in Scientific Discovery: A Chapter in
the Sociology of Science," cited above, and Thomas S. Kuhn, "Conservation
of Energy as an Example of Simultaneous Discovery," in Marshall Clagett,
ed., *Critical Problems in the History of Science* (University of Wisconsin
Press, Madison, 1959), pp. 321-356.

[29] David H. Bebeau, Hugh J. Miser, Dale E. Oyster, *The Estimated Effect
of a Soviet Atomic Attack on the U.S. in 1952 and 1954*, U.S. Air Force
Operations Analysis Office, Special Report No. 4 (October 18, 1950) ; a study
of the vulnerability of SAC bases in England done in 1950 by Hugh J. Miser
of the Air Force's Operations Analysis Office; and a twelve-volume WSEG
study dealing with SAC's capabilities done in 1950 reportedly touched
obliquely on problems of securing the strategic deterrent force. The short
supporting memoranda accompanying the Maddux and Craigie-Vandenberg
requests to RAND for studies also called attention in a preliminary way to
the dangers of an impending vulnerability of our strategic forces.

doctrines of the defense agencies reflected a clear appreciation prior to the base study of the need to achieve a secure deterrent capability. Some indication of how firmly entrenched were the traditional World War II categories of thinking and discourse of strategic air attack is given by the fact that even the Navy, in the bitter controversy with the Air Force over the B-36 in the late 1940's, showed no awareness of the potentially revolutionary implications of the vulnerability of aircraft on the ground to atomic attack. Striving to find ways to embarrass and discredit the Air Force, the Navy employed a wide variety of arguments against a further buildup of the strategic Air Force, but never raised the question of the vulnerability of aircraft on the ground to surprise atomic attack.[30] It is of course possible that the Air Force at some later date might have generated the policy changes on its own in the absence of the basing study, but unquestionably the costs (both in money terms and in terms of a lessened security of the deterrent) would have been much greater. It is also clear that it can become more and more difficult to reverse a policy course once vast sums of money have been committed and a program becomes firmly entrenched as part of the agency mission.

It remains to complete the story of the basing study's impact on Air Force policy. In September 1953 an important breakthrough occurred when the Air Staff Ad Hoc Committee reached agreement and endorsed the study's main recommendations. Its report concurred with the RAND findings that the programmed base system would be extremely vulnerable in the 1956-1961 period, and found the RAND-proposed overseas-refueling-base concept generally the most feasible

[30] See *Investigation of the B-36 Bomber Program*, Hearings before the Committee on Armed Services, U.S. House of Representatives, 81st Cong., 1st Sess., on H. Res. 234, August-October 1949.

approach to maintaining a high strike capability with reduced vulnerability. Of particular importance in the Ad Hoc Committee's report was the estimate in the Installations Section of the report that the overseas-refueling-base system would save at least one billion dollars over the programmed system in construction costs alone. This augured well for the disposition of the RAND study. For it held up the prospect of achieving a more secure capability *and actually saving money in the process.* One can imagine that a proposed innovation which *costs* an agency additional shares of the resources at its disposal might face additional obstacles.

In October, the issues went to the Air Force Council as General White had promised. There the Ad Hoc Committee appointed to evaluate the RAND study presented its report, after which Wohlstetter briefed the council for an hour and a half.

The Air Force Council continued its deliberations for some three weeks. Quite properly, only Air Force officers participated in these final deliberations of exactly how the RAND research would be translated into Air Force policy. In late October 1953 the council reached a decision on the following essential points: (1) The vulnerability of air-base facilities should be recognized in all Air Staff planning and action; (2) A hardening program should be initiated on critical facilities in overseas bases; (3) New overseas bases should be constructed to the specifications of ground-refueling functions; (4) Exceptions to these instructions will require special justification; and (5) Vulnerable stocks of materiel on overseas bases should be reduced. Shortly thereafter, in early November, the Air Force Chief of Staff ratified the council action.

Changes were quick to follow. The Air Force's plans no longer called for the deployment of the bulk of the strike force in overseas bases at the outbreak of hostilities. Instead, Operation Full-House, a new system employing the RAND proposed overseas-staging-base concept, was adopted.[31] The construction program on overseas bases was modified significantly; some critical facilities were dispersed and hardened. Runways at some new bases were made narrower so as to reduce construction costs and make possible the addition of more bases to the system (in keeping with the RAND suggestion that a larger number of cheaper bases would complicate the enemy's problem in launching a surprise attack). Measures designed to safeguard key personnel at overseas bases were put into effect. And plans for an expensive major depot in Alaska, on which it would be possible to do long-term maintenance of SAC bombers, were canceled.

In all, the base study had made several dozen specific suggestions. Some of these were adopted *in toto* by the Air Force, some were partially adopted, and some were never adopted. In addition, the Air Force initiated certain changes on its own that did not relate to any specific RAND suggestion, but that indirectly grew out of the Air Force's consideration of the problems raised by the basing study. The changes affected many parts of the Air Force hierarchy, including the functional command level and the Air Staff units responsible for war plans, logistics, installations, personnel and even the Directorate for Medical Requirements. The changes also were distributed over a time span: some went into effect almost immediately, while others were adopted only at various later points. Follow-up research to the basing study by Wholstetter and Hoffman led to further specific changes in Air Force policies and procedure. The "fail-safe"

---

[31] The official history of the Strategic Air Command (SAC) acknowledges an indebtedness to the basing study in the development of the Operation Full-House system.

procedure now employed by SAC, whereby SAC bombers are sent winging toward enemy targets on receipt of even ambiguous warning only to return automatically to their bases unless given an explicit Presidential order to proceed to target, was recommended in the next major study. The concept of the airborne alert also originated from the later RAND research. RAND suggestions on the importance of warning of enemy nuclear and thermonuclear attack, too, contributed to important changes in Air Force policies. Later RAND studies also pointed out certain problems of command and control under wartime conditions and the need for hardening ICBM sites which influenced Air Force policies.

In the end, the Air Force does not come out badly as a user of advice in the present case. It was neither uncritically receptive nor hostile to the suggestions. Although made uncomfortable by the findings of the RAND study, the Air Force recognized the need for innovation. And it acted with discretion and reasonable expedition to guard against the dangers brought to light in the study even though this involved substantial changes in its plans and operating procedures. The delay and opposition that developed was hardly surprising in view of the magnitude of some of the proposed changes.

Finally, it is appropriate to stress the important role for the advisor in facilitating innovation in the client agency. From its position as friendly critic, RAND injected heretical ideas into the Air Force hierarchy which stimulated needed change. The concept of an institutionalized critic, sufficiently independent to be provocative yet closely related so as to have access to information and the points of decision within the sponsor organization, will likely gain increasing acceptance in the future. Survival in an age of rapid change will require unusually flexible and adaptable organizational forms. A swiftly evolving science and technology have already had a visible impact on government organization in the area of national defense. As the "social fall-out" from developments in science and technology increasingly affect the operation of other government agencies, one can probably anticipate a heightened concern with innovative capacities and growing stress put on adaptable institutional arrangements.

CONCLUSION

This chapter has traced the evolution of a significant RAND study through its genesis in an Air Force request down to its reflection in Air Force policy changes some three years later. It is seen that the freedom of the researcher to reformulate the problem under investigation and to pursue a novel line of inquiry can have an important "payoff" for the policy maker. This type of long-range thinking is especially important in an era when technological developments are likely to bring changes in the political environment that are difficult to foresee on the basis of experience alone. This sort of function will continue to be important despite some tendencies in current military thinking to consider that the emergence of the missile and the nuclear warhead in operational form has produced a kind of technological plateau which reduces the urgency of forward planning and analysis. Rapid technological change has become a permanent state of affairs. The advent of a science-based technology in the twentieth century has destroyed the comfortable nineteenth-century notion that policy making was predominantly a matter of adjusting traditional thinking to gradual and incremental changes.

The example of the basing study also illustrates some interesting aspects of the question of how an interdisciplinary research effort like the RAND systems analysis is carried out. A large and amorphous "team" is not necessarily involved in the creation of a good systems analysis (though it may be on occasion). On the contrary, the present case suggests that small groups headed by one individual responsible for the end result, and operating informally to secure assistance from others when needed, more nearly represent a paradigm of the successful RAND systems-analysis project.

Further, this case study provides some useful glimpses into the intricate task of communicating the research product or advice to the client. Our analysis suggests that the position of the researcher or advisor outside the immediate governmental framework and his reputation for objectivity may contribute to effective utilization of the research or advice. In the present case, this factor operated to allow Wohlstetter to present his findings directly to General White, thus avoiding a permanent tie-up in the consideration of the analytic findings at a lower level. It served, further, to dissociate the RAND study from intra-agency conflict. This made it difficult to discredit the study on the basis that it sought to further the institutional and personal interests of a special group or sub-unit within the Air Force.

The decision-making process in "closed politics" is seen in the present case as containing certain built-in safeguards against arbitrary action based on uncritical deference to the scientific advisor. The RAND advisors faced a long and arduous struggle to bring their ideas to the attention of Air Force officers and keep the ideas alive in the face of inertia and skepticism and careful scrutiny at numerous points in the Air Force hierarchy. The decision maker does not emerge here in the passive and largely reactive role that Sir Charles Snow portrays in *Science and Government*. Nor is it evident, if the present case is at all typical, that the cardinal decisions are always or even usually made by a "handful of men." There is opportunity for laborious dissection of advisory recommendations at various policy levels. As in "open politics," the fragmentation of power and influence throughout many parts of the decision-making machinery means that action can be delayed or blocked or vetoed at a number of points. A handful of men can block action or decision with relative ease, but it is doubtful whether they can initiate it with the same ease. For favorable action on an advisory recommendation to occur, the advisor must normally persuade large numbers of people throughout the decision-making organization and foster something of a consensus among the interests affected by the recommendations. In short, "closed politics" seems to resemble "open politics" in a number of important respects.

Indeed, a last vital role we can discern for the advisory corporation like RAND is precisely to provide a link between the realm of "closed politics" and that broader public discussion of national-security issues which is essential to a flourishing democracy.[32] Allowing RAND and similar advisory institutions a semi-autonomous status outside the framework of the government keeps open channels of communication for debate and discussion of broad policy issues, and encourages exchanges of ideas and information between the "insiders" and the wider attentive public of journalists, publicists, academic scholars, and informed citizens. This type of institutional arrangement thus substantially reduces the danger that great issues will be decided by closed scien-

---

[32] For a discussion of some of Great Britain's difficulties resulting from lack of adequate linkage between the formal government establishment and outside critics and lay strategists (and informed citizens as well), see Laurence W. Martin, "The Market for Strategic Ideas in Britain," *American Political Science Review* 56:23-41 (March 1962). Martin concludes (p. 41): "It thus seems possible that, paradoxically, one of the more promising areas in which to look for improving the British defense machinery may lie outside the machine itself, in the encouragement of closer attention to questions of national security both by officially countenanced, specialized consultants, and by a wider attentive public."

tific politics or that the free play of policy debate will be choked off within a narrow self-contained elite.

As defense-policy choices become ever more complex, there can only be a growing need for some set of institutions to help serve as a bridge between the realm of "closed politics" and the larger public. This is needed not only on grounds of democratic theory—that is, to help counteract the tendencies toward technocracy inherent in the politics of the nuclear age. It is also needed because the formation and execution of a broad national-security policy adequate to the challenges of the nuclear age will require a high level of public understanding and will have to exploit a wide share of society's creative energies. The point is that "education of the wider public in questions of fundamental strategy seems particularly desirable, not only on grounds of democratic theory during an age when technology increasingly demands delegation of fateful decisions, but also because so many of the government's problems revolve around the question of what the public can be induced to support . . . Though there must be decision, it should be decision tempered by the freest possible flow of thought in the widest market compatible with national security: a limit likely to be generous on the more fundamental questions of strategy."[33] The advisory corporation like RAND can serve a vital function in tapping the intellectual capabilities of people outside the government for work on public problems and in generally promoting the public understanding necessary for the successful functioning of modern democratic government.

[33] *Ibid.*

► VII ◄

PROBLEMS AND PROSPECTS OF THE
NONPROFIT ADVISORY CORPORATION

On the basis of the RAND experience, can any general "guidelines" be inferred to govern the use of the nonprofit advisory corporation? Put in a slightly different way, to what extent does the RAND experience correspond to or throw light on the "proper" role for the advisory corporation? This chapter focuses on a set of broadly evaluative and analytical questions. The intention is to pull together some of the strands of our previous discussion into critical perspective. Much of what follows necessarily will be put forward in a tentative vein. The level of understanding of the organizational and psychological correlates of creative research does not permit many confident assertions. If the discussion at times resembles an agenda for further research, it seemed more useful at this stage to ask the right questions and to identify some of the pertinent problems than to attempt authoritative answers.

"IN-HOUSE" AND CONTRACT OPERATION

There is little need at the outset to dwell on the need for some kind of policy research in numerous areas of public policy. The accelerating scientific and technological complexity of the modern world, with its pervasive impact on the conduct of public affairs, has made some such "thinking," analysis, or planning function virtually indispensable for most government agencies. A number of questions arise, however, when one considers how such a function should be organized and related to the points of decision. It can be argued that a government agency, being responsible for policy in an area, should develop the capability for research and analysis as a normal part of its mission. The argument has considerable force. It is illusory to think that research or planning should be carried on only in special organizations (whether in or out of the government) specifically created for this purpose.[1] Every agency and individual, in whatever phase of policy making or operations, should show concern for planning, analysis of problems, reevaluating current practices, and informing his judgment before decisions are taken. As John M. Gaus has written: "Thinking and planning are a part of the task of every individual, if the policies that are evolved in order to realize the purposes of the organization are to be reflected in the day-to-day decisions."[2]

Yet it is clear that, as knowledge advances rapidly and policy choices become enormously complex, the normal staff competence within the agency may not fill all of the policy maker's needs. Many research functions will be best performed by highly trained specialists, teams of specialists, or separate research institutes. Should an agency attempt to develop special research competences of its own, or should

[1] Yet in 1918 it was still possible for the Haldane Committee in Britain to recommend a central Department of Research to serve the research needs of the entire government. Dahl and Lindblom comment that this was "a proposal that at the time seemed to exaggerate the role of research and by now seems like a quaint misunderstanding of its importance to every department, branch, and bureau of government." Robert A. Dahl and Charles E. Lindblom, *Politics, Economics, and Welfare* (Harper, New York, 1953), p. 79.

[2] "A Theory of Organization in Public Administration," in John M. Gaus, et al., *Frontiers of Public Administration* (Public Administration Clearing House, Chicago, 1936), p. 81.

it contract with an outside research institution? Stated in these general terms, the question does not admit of a very satisfactory answer. There are probably circumstances when each course of action is justified, and many occasions when one may serve as a useful supplement to the other. Having a research unit within the department clearly helps assure responsiveness to immediate needs, and may give the policy makers a better basis for dealing with many pressing problems. However, for the kind of long-range research that RAND seeks to perform it would usually seem desirable for an agency to contract with an outside institution. Systematic and imaginative policy research generally seems difficult to perform within an agency having responsibility for policy or operations. The pressure-ridden atmosphere of government agencies does not lend itself to the sort of intellectual "climate" conducive for creative policy research. And (what may amount to the same thing in practical terms) the individuals with the requisite research skills are difficult to recruit for direct government employment because they have become acclimated through years of training to an academic-type atmosphere.

The difficulties of performing genuinely useful "in-house" policy research can probably be mitigated to some extent by devising appropriate institutional arrangements within the government. For example, a separate research unit can be established and insulated as far as possible from the pressures affecting the rest of the agency. Geographical separation from an agency's headquarters may be another way by which the analytic unit can preserve some perspective and independence of viewpoint and operation. With leadership and personnel of stature the in-house research unit may occasionally achieve an autonomous status, and earn a reputation for quality work. Nevertheless, there are usually

243

enormous pressures working against the in-house analytic group which make its long-term prospects appear rather doubtful.[3] The tendency is strong to draw the group into current operations or to make it hew to a policy line. It may be inventive but often inventive only within very narrow limits. Frequently the in-house advisory unit's function becomes merely to provide ceremonial and ritualistic support for the policy maker in setting out on a new course of action or in bargaining with other agencies. The findings of studies referred to earlier suggest a marked tendency for research done in government agencies to reflect self-censorship on the part of the researchers and hesitancy to criticize the agency's program or policies.[4] Often the most useful policy research will be "heretical" in nature or sharply critical of current policy assumptions. And heresy does not flourish easily in an atmosphere in which there is a strong and understandable concern with agency-wide or government-wide policy coordination.[5]

Moreover, in a military bureaucracy with its characteristic and highly formal structure of authority there are special problems in fostering imaginative conceptual research and creative innovative activity. It is widely recognized that "out-

[3] In contrast, the government laboratory with a principally "technical" research mission seems to have a brighter future. This type of research operation seems to be easier to establish and to manage apart from an agency's policy concerns.

[4] Joseph W. Eaton, "Symbolic and Substantive Evaluative Research," *Administrative Science Quarterly* 6:421-442 (March 1962); Ashley L. Schiff, *Fire and Water: Scientific Heresy in the Forest Service* (Harvard University Press, Cambridge, Mass., 1962); and also Harry Eckstein's analysis of the "justificatory" attitude officials develop toward their own decisions in *The English Health Service* (Harvard University Press, Cambridge, Mass., 1958), ch. ix, "Epilogue: Planning in the Health Service," pp. 261-283.

[5] A similar problem is evidenced in private industry. When a firm wants objective advice on its policies and procedures, it often turns to an independent consulting organization that has no "stake" in the advice it provides.

siders" may often provide the impetus for change in program or doctrine.[6] Of the value of RAND in this regard, a distinguished former Air Force officer has remarked:

In the former case, the analytical and planning groups, their usefulness, I think, is to ventilate Government thinking about the longer term and the broader problems confronting one agency or another. In the case of RAND and the Air Force, for example, I know from my own experience in the Air Force that RAND was invaluable to us as staff officers in its capability and its ability to sit rather quietly out of the rut of the day-to-day operations in the Washington scene and think about what it is we are really up to here, where do we need to be ten years from now, and what should we start doing now to get ready for that, rather than to solve last Wednesday's problems, which all of us in the Pentagon have some tendency to do.[7]

In sum, for the long-range policy-research function there seem to be compelling reasons for often resorting to a contract arrangement with an outside institution. It would also seem clear that the sponsoring agency should avoid an overly restrictive specification of the work to be performed.[8] The contractor should report administratively at a fairly high

[6] See Chapter VIII below and the works cited in note 4.

[7] Gen. James McCormack (USAF, ret.), Vice President, Massachusetts Institute of Technology, in *Systems Development and Management*, Hearings before a Subcommittee of the Committee on Government Operations, U.S. House of Representatives, 87th Cong., 2nd Sess. (the Holifield Subcommittee Hearings), part 4, August 1962, p. 1537.

[8] Ronald W. Clark's observation regarding the British experience with operations research in World War II is pertinent: ". . . the successful Operational Research officer would have to have access to secret material at the highest level; for him, quite as clearly as for the highest commander in the field, it was impossible to work properly without full information. But at the same time it was useless to give an Operational Research Group anything in the nature of daily orders or set objectives. Its members had to understand both operations and operational urgency; but they had also to look in from the outside, with the objective curiosity of the scientist, deciding for themselves what it was most necessary to do next. 'The secret of science,' Tizzard once said, 'is to ask the right question, and it is the choice of the problem more than anything else that marks the man of genius in the scientific world.'" *The Rise of the Boffins* (Phoenix House Ltd., London, 1962), p. 214.

level in the client agency to lend some prestige to the operation and to help assure that the research results will receive a serious hearing. But the contractor probably should not be administratively attached to the highest levels. There seem to be greater difficulties in gaining access to information and in preserving independence of operation when the research institution works directly for the highest political levels.

The considerations arguing for an outside contractor perhaps would not operate so forcefully with respect to a systems management or program supervision function. These do not require the same high level of conceptualization and independence of viewpoint as the long-range "think" function. Nor would those advisory duties geared to the "short-run" and designed to be fairly responsive to an agency's immediate concerns necessarily require location outside the framework of the government. The kind of advisory task performed by ANSER, for example, might appear as a logical choice for an in-house operation.

Some further clarifying remarks seem in order at this point. First, the above comments should not be taken as a doctrinaire effort to identify a certain type of research assignment exclusively with a certain organizational setting. There is some virtue in attempting to assess the strengths and weaknesses of different types of organizations for performing various research functions, but an overconcern with logical neatness or administrative tidiness would miss the essential point. The principal objective of a system of advisory services is not a clear structuring of roles but rather a guarantee that significant problems and policy alternatives are constantly thrown up for the policy maker's consideration. An effective system probably should include advice from a number of different points, some interested and some disinterested, the advice sometimes parochial and sometimes broad in scope, some sources concerned with the long-run and some with the issues of the moment. In short, a "mix" of institutions and outlooks helps assure that all important points of view are represented and that significant problems are not neglected.

Second, any contract operation presupposes a sufficient level of research competence within the government agency so that contract studies may be properly evaluated. Without this competence public accountability for policy is jeopardized. No one doubts that policy decisions should remain the responsibility of government officials. But "policy" here can only mean those issues which government officials consider important enough or feel strongly enough about to decide for themselves.[9] If they lack the competence to evaluate studies or to understand what warrants a policy decision on their part, decision-making power will gravitate to the knowledgeable outside experts.[10]

Third, there is a range of basic management functions respecting the government's research and development program which should never be contracted out. This injunction applies equally well to policy research as to the letting of contracts for military hardware or for the supervision of development projects (though it is particularly important in the latter cases). "Decisions concerning the types of work to be undertaken, when, by whom, and at what cost must be made by full-time government officials clearly responsible to the President and to the Congress."[11] RAND, for example, should never exercise the function of evaluating the work of

[9] For an elaboration of this point, see Don K. Price, "The Scientific Establishment," *Proceedings of the American Philosophical Society* 106:243-245 (June 1962).

[10] See then Budget Director David E. Bell's testimony in Holifield Subcommittee Hearings, part 1, esp. pp. 43-49, for an interesting discussion of this point.

[11] Bell Report, in Holifield Subcommittee Hearings, part 1, p. 214.

other advisory corporations or of deciding when they should be awarded contracts for studies. Further, the lines between the agency and contractor must be kept clear enough so that there is no blurring of responsibility for these basic management functions.

Finally, there are several general considerations that agencies sponsoring contract-research operations might bear in mind. It seems desirable for the government to keep in business a substantial community of advisory institutions. Just as there is a need for technological base, it would seem important to have something like an "advice base" for government agencies. A larger number of smaller institutions is probably preferable to a small number of large institutions because this maintains a broader and more flexible base of capabilities. By parceling out contracts to different contractors, and by keeping enough research contractors in business to permit some competition between them, the government can also prevent any one or any small group of institutions from monopolizing the avenues of access to persons in authority.

The question of a research institution's size may be an important issue in its own right. There is some reason to suppose that, as a research organization grows larger, bureaucratic rigidities and additional administrative controls may develop which impede the research effort.[12] How large is

12 See Chapter IV above and the studies cited in note 43. Also of some interest in this context are remarks made in 1962 by James M. Bridges, then Director of the Office of Electronics, Department of Defense Research and Engineering. Bridges was quoted as saying that the productivity of electronics engineers and scientists working for certain nonprofit institutions is "approximately 10 times greater" than that of their counterparts in private industry. Bridges, relating his remarks specifically to the Massachusetts Institute of Technology, gave two reasons for his belief: (1) the MIT research groups were "small" and (2) were composed of "highly creative" people. No criteria for making the judgment are given. Speaking in broad terms about bureaucracy's significance in modern society, Max Weber made a suggestive observa-

"large" for the RAND-type operation? Should there be a policy of no growth? Slow growth? One can perhaps hazard some cautious inferences, but in the main this is an area of ignorance badly in need of systematic research. Rapid growth, at one extreme, probably is inconsistent with an effective policy-research effort. Rapid growth generally has an unsettling effect upon an organization, a circumstance that could disrupt the research environment. Yet at the other extreme no growth seems an unpalatable alternative. A certain amount of growth may be necessary to maintain an organization's vitality.[13] Thus an arbitrary "ceiling" on the size of the advisory institution would appear unwise.[14] Moreover, there are surely some important economies of scale available to the research organization of a certain size (and possibly some threshold below which an organization lacks sufficient skills to generate a productive research effort). Nevertheless, it seems likely that there are hazards to increasing size and that beyond some broad limit further growth may pose significant problems for both the sponsoring agency and the research institution's management.

PROFIT AND NONPROFIT CONTRACTORS

A major strength of the system evolved in the United States to support various kinds of research work is its versatility and

tion that is also worth pondering in this context: "Only by reversion to . . . small-scale organization would it be possible to any considerable extent to escape its [bureaucracy's] influence." *The Theory of Social and Economic Organization* (Oxford University Press, New York, 1947), p. 338.

[13] The problem of an organization with little or no prospects for growth was touched on in an exchange between Dr. Stanley Lawwill, head of ANSER, and Mr. Herbert Roback, Staff Administrator, in the Holifield Subcommittee Hearings, part 3, pp. 1083-1084.

[14] A "ceiling" on RAND's size has been discussed in the Air Force for a number of years, but as yet no formal policy to this effect has been adopted. The "ceiling" notion can be variously interpreted to mean either a fixed manpower level or a fixed dollar amount of support. Either way such a policy could have some serious consequences for an organization like RAND.

adaptability. No one specific institution or set of institutions is indispensable. The government can turn to many different sources to accomplish scientific and analytic work needed for public purposes. There is a dynamic quality and a certain looseness around the edges to the system. Various research and development tasks could be, and have been, performed by different kinds of organizations. Thus the system has never hardened into any inflexible organizational pattern. In general, it seems important to maintain a measure of looseness and fluidity in the system with capabilities overlapping to some extent.

However, it is appropriate in view of recent controversies to discuss some of the special strengths and weaknesses of different kinds of research organizations. In particular, there has been an intensive debate in recent years over the proper roles of profit and nonprofit organizations in defense research and development work. Industry, and to some extent the non-industrial profit-making research entities as well, have viewed the nonprofit institutions as dangerous competitors operating under the cloak of tax immunities.

In reality, much of the debate has been wide of the mark. Industry's fears have been exaggerated, as the bulk of defense contracting has remained with industrial concerns.[15] The use of nonprofit institutions for analytical services hardly constitutes a direct threat to industry in regard to the manufacture and sale of products to the federal government. Even the systems-management nonprofit corporations like Aerospace and MITRE have been careful not to overstep

[15] See *Federal Funds for Science XI*, Fiscal Years 1961, 1962 and 1963, National Science Foundation, NSF 63-11 (January 1963), pp. 17-22. 79 percent of extramural R & D obligations by the federal government for fiscal year 1962 went to profit-making organizations, and nearly two-thirds of all federal obligations for research and development for fiscal year 1962 went to profit-making organizations.

their grants of authority and tread on the toes of industry.[16] Industry's position seems secure. In almost all influential sectors of society it is agreed that industry should remain the dominant developer and producer of goods for sale to the federal government. To be sure, some small percentage of development and production work is done in government laboratories and nonprofit installations. Some believe it would be in the public interest to expand the percentage somewhat. But there is no influential body of opinion in the country which seeks to put the government or nonprofit installations into the position of making military or space systems or other production items on any substantial scale. Industry's position is also underwritten by strong political support from the Congress.

As for the field of "paper-pure" analytical services, it cannot be said that industry has any traditional expectations for a large share of the business. The business of providing technical-policy advice to a defense agency is *new* and has no clear precedent. If anything, the nonprofit corporation in the relatively short history of institutionalized advisory processes has become the more "traditional" setting for this work. In any event, industry is logically in a minor role here because the analytic work in part involves objective evaluations

---

[16] See the testimony of Ivan C. Getting, President of Aerospace, in Holifield Subcommittee Hearings, part 3, pp. 978-979. Getting observes: "I think that the general reaction that I have found is that industry will always be suspicious a little bit. They always resent getting technical direction. But on the whole, they would much rather get it from a competent organization like Aerospace which has no irons in the fire, than get no technical direction and be in confusion about what they should be doing and how. I have found no real opposition from industry . . . I find that as long as we do not raid them unscrupulously of good people, as long as we don't pay salaries that are higher than theirs, so long as we don't take credit away from them, so long as we allow them to carry their heads on their shoulders and say 'We did it, we have smart people, too,' that our community relations with industry will remain good."

of the capabilities of systems that industry might be called on to build. It is sharply at variance with standards of ethics and propriety in government-contractor relations for any firm to be in the position of advisor to a government agency at the same time it stands to benefit from the possible award of development or procurement contracts from the agency. Thus, for example, General Electric's TEMPO would be in an anomalous position if it were to act as a military service's regular advisor in view of the substantial volume of development and production work that General Electric performs for the services. Nor would an RCA-affiliated group normally be a likely choice for advising NASA on the potentialities of alternative communications satellite systems. One may question whether NASA's relationship with Bellcomm, Inc., an American Telephone and Telegraph affiliate, will prove viable for precisely these reasons.[17] There may be some point in having industry-affiliated advisory units as a means to keep industry abreast of current strategic thinking and to relate planning for the development of future systems to changing military needs. The industry advisory unit may also play some role from time to time in performing *ad hoc* studies for government clients in areas where it has a special competence. But it seems highly unrealistic to expect the industry-affiliated advisory institute to play a major role in the defense-analysis field.

Industry's extreme hostility toward the nonprofit organizations, including the advisory organizations whose work is seemingly so unrelated to industry's main concerns, can

---

[17] For a brief description of the NASA-Bellcomm relationship, see Holifield Subcommittee Hearings, part 5, pp. 1751-1776. Some kind of "hardware ban" on AT & T, such as the Air Force imposed on STL and the Thompson-Ramo-Wooldridge Company, might allay suspicions in this case. But as in the STL situation the hardware ban would probably be only a palliative and really satisfactory to no one. See note 14 of Ch. IV above.

perhaps be attributed in part to a fear that organizations like RAND have helped to fashion a climate of opinion in which proposals for new weapons systems are increasingly subjected to critical scrutiny. The nation's security could be seen as being advanced in many cases by *not* developing new systems, an assumption sharply at odds with traditional defense-industry views. Industry would likely feel in a better position to influence the frame of reference of the strategic debate if the advisory institutions had closer industrial ties.

There is more justice to the complaints of the profit-making research organizations with no direct industrial ties.[18] These organizations normally engage in a variety of research and advisory tasks for numerous government and non-government clients. Their number includes such organizations as Arthur D. Little & Company, Planning Research Corporation, E. H. Plesset & Associates, the Council for Economic and Industrial Research (CEIR), and also to some extent the management consulting firms like Booz, Allen, and Hamilton.[19] Their essential complaint is that their nonprofit competitors are subsidized by the tax advantages accruing to nonprofit status, hence their own costs are higher, and they do not get a fair share of the business. The argument in these narrow terms

[18] "Profit making" here means those research organizations which are organized for financial gain, distribute shares to stockholders, and can declare bonuses which inure to the benefit of individuals. "No direct industrial ties" means that these organizations normally are not engaged in the business of manufacturing and selling goods on the market; they are consultant firms, not producer firms.

[19] Arthur D. Little & Company seems to have assumed a position of leadership among this class of organizations, and become something of a spokesman for their interests. Helge Holst, Arthur D. Little treasurer and corporate counsel, took primary responsibility for organizing the group that produced the "Industry Bell Report." See the Holifield Subcommittee Hearings, part 1, pp. 79-121, 339-402. Some of these profit-making research entities, it should be recalled, arose when entrepreneurial researchers broke off from RAND and formed their own corporations. See note 37 of Ch. III.

is correct, but it hardly seems reasonable for government agencies to pay more for a service when they can often get it for less. Moreover, the fact that the profit-making research organizations have a divided industrial and government clientele may be significant. Conflict-of-interest situations are more likely to occur with organizations having a partly industrial clientele than with organizations performing advisory services only for the government. This may help explain why the profit research groups have been used less in the role of "permanent" advisor to a particular government client but have instead performed various *ad hoc* tasks for different government clients. The need for analytical services, in any case, is such that there is scope for both profit and nonprofit organizations. If a larger number of genuinely competent profit advisors develops, it can be anticipated that they will get a share of the business.

What role is there, if any, for the university in the defense-analysis picture? The university community clearly contains a deep reservoir of skills useful in defense analysis or other technical advisory work. In some ways, the academic community seems an ideal skill base to employ for contract advisory work. There are some university research centers, such as MIT's Center for International Studies, which could serve as a precedent for an expanded university role. In general, however, it seems not in the best interests of either the academic community or society at large for universities to take a prominent role in performing policy research on contract for government agencies. The university has traditionally been a source of independent and creative thinking unencumbered by obligations to a client and freed from the necessity of proving "useful" in an immediate sense. This traditional independence might be compromised if universities were to move directly into the business of advising government clients on a significant scale. Already some serious issues have been raised by the extent of federal support for university science with respect to a possible erosion of academic freedom of inquiry.[20] Whatever the merits of these specific issues, there can be little doubt that assuming the status of confidential advisor to government agencies would pose significant new problems for the university-government relationship.

Moreover, in the defense-analysis area there is the added factor of classified materials which require careful handling and strict clearance procedures. Although some universities have worked on classified research projects without compromising their quest for pure knowledge, it remains true that security restrictions are in general at odds with the spirit of open discussion and unrestricted scholarship vital to the university. Any further incursion of classified research into the university environment, therefore, must be viewed with suspicion. One must also view the matter from the perspective of government officials concerned with safeguarding sensitive information. Clearly, it is difficult to locate a secure classified facility in an institutional setting where all the mores and traditions motivate individuals to be open and candid in discussing their work.

The more normal role for the university in this area, rather, would seem to lie in its traditional role of generating basic intellectual capital and training new generations of scholars

---

[20] See, inter alia, Harold Orlans, *The Effects of Federal Programs on Higher Education* (The Brookings Institution, Washington, D.C., 1962); Charles V. Kidd, *American Universities and Federal Research* (Harvard University Press, Cambridge, Mass., 1959); James McCormack and Vincent Fulmer, "Federal Sponsorship of University Research," in American Assembly, *The Federal Government and Higher Education* (Prentice-Hall, Englewood Cliffs, N.J., 1960), ch. iii; and J. Stefan Dupré and Sanford A. Lakoff, *Science and the Nation* (Prentice-Hall, Englewood Cliffs, N.J., 1962), ch. iii, "Universities and Government."

(some of whom will staff the contract advisory organizations). Where special skills are needed that are available only within the university, the device of the individual consultant can be employed. Thus the government can have the services of special academic talents on a temporary basis, while leaving the university environment essentially unaffected. Through such contacts as the consultantship, scholars also can become interested in studying national-security problems as a part of their normal professional interests. This type of interest freely given can be regarded as a "bonus" for the government, and represents a potentially valuable addition to the creative attention devoted to public problems.

An important qualification, however, must be entered to the above comments on the university's role. The university's role in this area should probably become much greater if, for one reason or another, the advisory institutions with no formal academic ties encountered growing difficulties in recruiting qualified staff. One can imagine this happening, for example, if a sponsoring agency imposed excessive bureaucratic and political controls on the contractor. The advisory institution can attract good people and can be an exciting place to work only if there is a large measure of intellectual "elbow room." If unwise administrative policies should choke off this quality, then a university setting might appear a more secure institutional base to draw on qualified personnel and to resist client pressures.

CONFLICT-OF-INTEREST

In general, the whole area of potential conflicts-of-interest of individuals and organizations in the R & D field is in an early stage with respect to developing accepted standards of conduct—unlike other fields, such as the law or medicine, where there are long-established standards of conduct.[21] Many ill-defined areas exist which are not fully understood, and only recently have efforts been initiated to clarify the archaic statutory provisions (most of which antedate the scientific revolution of the twentieth century) dealing with conflict-of-interest situations. The position of organizations like RAND, in particular, seems to be among the least understood aspects of the situation. What is the conflict-of-interest problem that an organization like RAND faces?

A first aspect of the problem concerns the avoidance of potential conflicts-of-interest of what the Bell Report has termed an "organizational" nature.[22] This concept is not well-defined, but perhaps some idea of what it implies may be obtained by referring back to a phase of RAND's history. We saw that for a time RAND considered, then abandoned, the idea of spinning off the training activity that eventually became a separate nonprofit corporation—the System Development Corporation—as a profit-making concern loosely attached to RAND. If the Air Force and RAND had been imprudent enough to allow such an arrangement to materialize, an "organizational" conflict-of-interest would have been created. For RAND would have been in a position to give advice to the Air Force on matters that could involve the granting of a research, development, or production contract to an organization in which RAND had a financial interest. Or, to take another example, had RAND remained a part of the Douglas Aircraft Company, a similar sort of conflict-of-interest would almost certainly have arisen sooner or later. Still another example of a conflict-of-interest relating to an

21 Bell Report, in Holifield Subcommittee Hearings, part 1, p. 225. See also New Rork City Bar Association, *Conflict of Interest and the Federal Service* (Harvard University Press, Cambridge, Mass., 1960).

22 Bell Report, in Holifield Subcommittee Hearings, part 1, pp. 225-227.

organization rather than to an individual would be the situation resulting if RAND were to invest corporate assets in stock of a firm that stood to benefit from the award of defense contracts.[23] An "organizational" conflict-of-interest, then, refers to an organization placing itself in a position which inherently raises conflict problems whatever the motivation or intentions of individual officials. It is apparent that both the advisory institution and the sponsoring agency should be on guard against such situations and, wherever possible, clear standards of acceptable conduct should be formulated.

The second salient aspect of RAND's conflict-of-interest problem involves the observance and enforcement of standards of acceptable personal conduct on the part of individual RAND staff members, particularly in their dealings with industry and the government. This applies to all levels of the RAND staff, including the Board of Trustees. Some concern has been felt that "interlocking directorates" on the boards of various research establishments could constitute unhealthy avenues of influence and pose possible conflict-of-interest situations. The Bell Report noted that "there is a significant tendency to have on the boards of trustees and directors of the major universities, not-for-profit and profit establishments engaged in Federal research and development work representatives of other institutions involved in such work."[24] The report went on to cite several possible justifications for this practice, but concluded with the following warning: ". . . we see the clear possibility of conflict-of-interest situations

[23] A fanciful misrepresentation of RAND's conflict-of-interest situation is found in Saul Friedman, "The RAND Corporation and Our Policy Makers," *Atlantic Monthly* 212:61-68 (September 1963). Friedman asserts (p. 64): "Rand makes most of its money by charging the air force 6 percent of the estimated costs of the contracts which the air force lets to private industry as a result of Rand's work." RAND's fixed fee is 6 percent of the dollar amount of the contract with the Air Force or any other client.

[24] Bell Report, in Holifield Subcommittee Hearings, part 1, p. 224.

developing through such common directorships that might be harmful to the public interest." Clearly, it would seem important for an organization like RAND (as well as other R & D establishments) to develop some guidelines for trustees whose membership on other boards might affect their actions as RAND trustees.

Individual staff members might become involved in conflicts-of-interest in a number of ways—some of which can be clearly recognized and some of which are difficult to foresee. Let us first consider the relation of RAND staff members to industry. A RAND study team, for example, might urge the Air Force to adopt a new kind of weapons system, knowing that a certain firm is the only company producing the system or the likely company to receive a contract. On the basis of inside information, the individuals might speculate in the firm's stock—or, an even more obvious conflict, might form a direct business association with the firm.[25] Another sort of conflict-of-interest involving contacts with industry is the following: individuals could attempt to use for personal gain proprietary information obtained from one firm in their dealings with a second firm. How a situation like this might arise is apparent and requires no elaboration.

In many of RAND's dealings with industry, however, there is a kind of built-in protective mechanism that deserves mention. The intensely competitive character of the industrial world serves as a powerful incentive for organizations like RAND to guard against efforts at personal gain by staff members in their dealings with industry. If a RAND staff member

[25] Referring back to an earlier discussion, we saw that RAND may become involved in recommending an expansion of the use of a metal such as beryllium. Now had the RAND researchers involved in this project sought to capitalize on their knowledge by becoming associated with one of the few firms mining beryllium or by speculating in beryllium stock, a conflict-of-interest of the sort we are discussing might have been created.

sought to aid one firm in return for some favor or emolument, the competing firms would likely find out sooner or later and would raise a severely damaging political outcry. RAND would also doubtless find its access to industrial proprietary information quickly dried up. The absence of any serious accusation of bias or partiality in this respect, despite bitter criticism of the "nonprofits" by industry, attests the care and prudence of RAND management and that of other nonprofit institutions. This self-regulating mechanism is far from infallible, of course, as there are a number of areas of defense research and development work where competition is largely absent. A different sort of problem faced by IDA and not by RAND is the presence of employees "on loan" from industry on IDA's staff. This practice would seem to pose some potential conflict-of-interest difficulties.[26]

Conflict-of-interest situations involving the relations of individual RAND people with government officials provide a related, and perhaps even more difficult to define, area where standards of acceptable conduct must be developed. As previously noted, large numbers of RAND people serve on a wide variety of government advisory committees and as consultants to a number of government agencies. It is clear that these advisory duties could place the RAND staff member in a position where he could advance RAND's interests (and indirectly his own) through his close contact with government officials. Given the intimate partnership between private institutions and the government in the R & D effort, however, it is equally clear that such contacts must and will continue on a significant scale. The problem is not *whether* there should be extensive contacts between government and nongovernment persons which make conflicts-of-interest pos-

---

[26] See the testimony of former IDA President Richard Bissell, Holifield Subcommittee Hearings, part 2, pp. 626-627, for a discussion of this problem.

sible—the nature of the system demands that there must be these contacts—but *how* we can best regulate these contacts to minimize the likelihood of unhealthy situations while not interfering substantially with the progress of the research work. There is no blinking the fact that sometimes the objective of minimizing potential conflicts-of-interest and that of achieving maximum progress in the scientific work will conflict, and one will have to be given priority over the other.

The President's memorandum of 9 February 1962 to the heads of the executive departments and agencies, "Preventing Conflicts of Interest on the Part of Advisors and Consultants to the Government," takes a first step toward defining standards of conduct applicable to advisory and consultant relationships.[27] Although much of the memorandum is not applicable to RAND's situation, some of its standards do apply to phases of RAND's activities and, in general, the spirit of the memorandum was intended to encourage organizations like RAND to review their own conflict-of-interest policies.

For the first time, the President's memorandum provided that section 434 of Title 18 of the U.S. Code would be interpreted to include nonprofit organizations and educational institutions within the meaning of "business entity" as used in the section.[28] Furthermore, the memorandum broadened the interpretation of the phrase "the transaction of business" to include "the rendering of advice which will have a direct and predictable effect upon the interests" of any entity in

[27] Reprinted in Holifield Subcommittee Hearings, part 1, pp. 34-41.

[28] Section 434, one of the six statutory provisions outlawing conflicts-of-interest, prohibits government employees (and consultants and advisors while actively employed) from transacting business as a part of their government duties with any business entity in which they have a financial interest. See New York City Bar Association, *Conflict of Interest and the Federal Service*, pp. 42-44.

which the advisor or consultant has a business or financial interest.[29] The broad standard was laid down, in other words, that no individual serving as an advisor or consultant should give advice on an issue directly affecting the interests of any private organization which he serves, whether it is a university, nonprofit corporation, or business firm.

Another standard laid down in the President's memorandum that applies to RAND people serving as consultants and advisors to the government is a prohibition against stock speculation using inside information gained through the advisory work. The major means envisaged in the memorandum to enforce this prohibition is a requirement for the consultant or advisor to disclose the essentials of his employment and financial interests upon appointment.[30] Since RAND as a whole is a group of advisors to the government, it would seem desirable to extend such a prohibition to all RAND staff members (and other nonprofit advisory personnel) as well as to those serving as intermittent advisors and consultants.

Although the standards developed in the President's memorandum are far from exhaustive or entirely unambiguous, the memorandum nonetheless has served a valuable purpose. It has taken an important first step toward clarifying a difficult and uncertain area. Even apart from its substantive

[29] President's Memorandum, February 9, 1962, in Holifield Subcommittee, part 1, p. 37.

[30] *Ibid.*, pp. 39-40. The memorandum specifies that the disclosure statement "indicate the names of all the companies, firms, research organizations and educational institutions which he is serving as employee, officer, member, director, or consultant, and the companies in which he has any other financial interest, such as the ownership of securities or other interests which have a significant financial value . . . Each statement of interest should be forwarded to the chief legal officer of the department or agency concerned, for information and for advice as to possible conflict of interest. In addition, each statement should be reviewed by those persons responsible for the employment of consultants and advisors to assist them in applying the criteria for disqualification discussed in this memorandum as set forth above . . ."

merits, the memorandum has been useful in stimulating government agencies and private organizations to think seriously about the issue.

RAND has responded to the Presidential memorandum by undertaking a review of its conflict-of-interest policies, and in particular scrutinizing all cases of RAND personnel serving in an advisory or consultant capacity to any government agency.[31] RAND has also made it a policy that no staff member should accept any new position as an advisor or consultant to the government, or renew any such existing arrangement, until the front office has reviewed the matter in light of the President's memorandum and any subsequent implementing directives from the Department of Defense. In general, RAND management since the end of the 1950's has grown increasingly sensitive to possible conflict-of-interest dangers owing largely to a heightened congressional interest in the subject.

In 1960, RAND adopted a partial disclosure requirement for key employees to guard against possible stock speculation using inside information gained in the course of their work. Department heads were asked to disclose any substantial stockholdings which might involve a conflict or the appearance of a conflict with any of their current work. The department heads were also to apply similar requirements to their department staffs, and to bring any pertinent information to top management's attention. In practice, no serious abuses seem to have occurred under this system, although it is possible that RAND could formalize and further improve its policies in this area. A tightening up of internal regulations in this sphere will almost certainly become necessary as

[31] Memorandum to RAND Research Staff from J. R. Goldstein, "Conflicts of Interest," M-1455, February 16, 1962. For information on the steps RAND has taken to avoid conflicts, I am indebted to interviews with corporation officers, J. R. Goldstein, Vice President, and S. P. Jeffries, Secretary.

RAND grows larger and it is no longer possible to review difficult situations informally on a case-by-case basis. The spirit, if not the letter, of the prohibition against stock speculation contained in the President's memorandum of February 1962 should provide the impetus for all organizations of the RAND type to adopt appropriate safeguards.

A neglected area at present, which needs careful management surveillance, is that of consultant relationships between the employees of different nonprofit institutions. On some occasions, for example, RAND staff members have acted as consultants for a sister group like IDA on matters related to their RAND work and have received consulting fees for this service. This practice would seem improper. RAND and other advisory corporations should take steps to prevent employees from receiving additional fees of this kind. In the absence of appropriate safeguards, one can imagine a situation developing where various advisors will regularly supplement each other's income by exchanging consulting invitations. Extensive staff contacts among the advisory organizations must and should continue. But they should be on a voluntary basis with no one in a position to receive supplemental income for making available to other government contractors the knowledge he has acquired in his own government contract work. It should be noted that this sort of potential conflict-of-interest is by no means unique to the nonprofit advisory corporations. Numerous institutions, profit and nonprofit, face a similar need to develop appropriate standards of conduct in this area.[32] We see here just one of many in-

[32] The Atomic Energy Commission, for example, has permitted employees of its university-managed contract laboratories to accept consulting invitations from industrial firms to comment on research they have done (or are doing) under government contract. Industry may then charge off these costs to their own government contracts. In effect, this amounts to double compensation for the same work.

stances of unusual conflict-of-interest situations arising in the new partnership between government and science.

Yet on balance it does not appear that the conflict-of-interest issue poses inherently insoluble difficulties for the RAND-type institutional arrangement. On the contrary, RAND's experience suggests that careful management on the part of both the government and the advisory organization can control most potential conflicts-of-interest, and provide a generally suspicion-free working relationship in the public interest. As more experience is gained, standards of conduct are likely to be further refined and more effective ways found to deal with remaining problem areas. Indeed, it is doubtful whether any alternative arrangement would offer many additional guarantees. Even to incorporate all private advisors formally into the government would not eliminate all possible conflicts-of-interest. With the complexity of modern government, personal and institutional self-interest can pose threats to the public interest even within the formal government framework. Vigilance, good judgment, and constant attention are required under any organizational arrangement. In the end, beyond any formal standards of conduct or formal procedures, the sensitive conscience of individuals in the government and in private organizations remains a vital safeguard.

HOW MUCH INDEPENDENCE?

The importance of "independence" for the advisory organization has been implicit in much of our previous discussion. What does this concept mean? In what sense can and should the advisory organization be "independent" of the agency or agencies it serves?

The first step toward understanding the "freedom" and "independence" of the advisory organization is to recognize that these cannot be absolute. The advisory organizations cannot be grouped neatly into mutually exclusive "independent" or "nonindependent" categories. Each to some extent operates under various constraints, formal and informal, which interfere with its independence to greater or lesser degrees. Like the individual in a complex industrial society, the advisory organization searches for freedom between and within shifting networks of pressures, authority structures, group loyalties, and other constraints which pattern its existence.[33] The organization like RAND, in brief, must find its freedom and independence in the interstices of the administrative and political controls under which its operates. Some sacrifice of independence is perhaps one of the prices an advisory organization must pay for having a voice in policy formation. As a leading student of government organization remarks: "So-called 'independent' corporations have paid a high price for their independence by being deprived of an effective voice in the formulation of national policy."[34]

Recognizing realistic limits to the advisory organization's independence, however, should not lead one to conclude that degrees of independence are not obtainable or desirable. In actuality, RAND's prominence in the defense-analysis field is certainly related to the surprisingly high degree of independence that it has enjoyed over the years. What is striking about RAND's experience, indeed, is the extent to which it

[33] See the brilliant essay by Louis Hartz, "Democracy: Image and Reality," in William N. Chambers and Robert H. Salisbury, eds., *Democracy Today: Problems and Prospects* (Collier Books, New York, 1962), pp. 25-44, for the argument that freedom in our complex industrial society must be found in the interstices between the various group memberships and other forces that constrain the individual, and see also Barrington Moore, *Political Power and Social Theory* (Harvard University Press, Cambridge, Mass., 1958), ch. vi, "Reflections on Conformity in Industrial Society," pp. 179-196.

[34] Harold Seidman, "The Government Corporation in the United States," *Public Administration*, Summer 1959, p. 111.

has been able to avoid the role of merely providing cere-monial and ritualistic support for a client's current interests.

RAND has enjoyed substantial freedom of action in a number of spheres, but several areas have been particularly important. The first of these might be termed "program independence." More than any of the sister advisory corporations, RAND has customarily been able to initiate studies of its own choosing, reformulate Air Force suggestions for research projects, and generally to make its own decisions about the kind of work that would be most valuable to the client.[35] This kind of program independence has enabled RAND to maintain a substantial degree of internal freedom, and has been a vital asset in attracting qualified staff. The need for program independence must be considered the crucial aspect of the whole question: nothing else matters much unless the advisory organization is allowed considerable freedom of action in regard to the staffing, planning, and execution of the research program. An organization like RAND might survive if forced to add to its charter a pro-vision specifying that all corporate assets would devolve upon the government in the event of dissolution. But it would not survive as the same kind of organization if there were an erosion of its privileges to control the direction of the research effort.

A second critical area might be termed "publication inde-pendence." It is inevitable that any defense advisory orga-

[35] This assertion is difficult to document with any precision, but it is the almost unanimous opinion of knowledgeable persons "in the trade." Some documentation may be found in the testimonies before the Holifield Sub-committee Hearings. See the *Hearings*, part 2, pp. 629-630 (concerning IDA), and part 3, pp. 1079-1080 (concerning ANSER). Also, the original RAC contract document provided that 10 percent of RAC's workload could consist of self-initiated studies. The comparable figure for RAND was 37 percent (see Chapter V) with the probability that this percentage figure did not accurately reflect the full extent of RAND's actual program independence.

nization will be subject to certain controls over the publication and distribution of reports (and over such matters as the freedom of the research staff to give outside speeches on matters of client interest). Some controls of this sort are probably legitimate means to protect valid client interests. But these controls are also among the most dangerous and subject to abuse of the management devices at the client's disposal. Whether or not these controls are exercised with restraint and good judgment can make a profound difference. A particular sore point is often the client's readiness to attach security classifications to material for policy reasons. The Air Force, for the most part, deserves considerable credit for its administration of Project RAND in this sphere. The Air Force has granted RAND wide discretion to publish unclassified materials, permitted considerable latitude in the distribution of classified writings to cleared personnel in other government agencies, and generally allowed RAND staff members to speak their minds freely even on subjects of official concern to the Air Force. It is perhaps not too much to say that the relationship between RAND and the Air Force here presents something of a paradigm for a mutually beneficial sponsor-advisor arrangement.

In general, it would appear that RAND's sister advisory organizations could benefit from greater independence in the program and publications areas. Such changes would better equip organizations like the Institute for Defense Analyses (IDA), Research Analysis Corporation (RAC), and the Navy's Center for Naval Analysis to perform a creative policy-research function.[36] Even given a much greater under-

---

[36] It should be recalled, however, that some phases of the operations of RAC, IDA, and the Center for Naval Analysis are designed to provide immediate operational assistance. As such they do not require the independent setting needed for a creative policy research function.

standing of novel research activities in government circles, however, one cannot be very optimistic that these organizations could ever achieve the same unusual degree of independence that RAND has enjoyed. RAND's unusual independence, as we have seen, was partly the result of historical accident. It seems unlikely that a confluence of favorable circumstances such as aided RAND's development can be counted on to occur with any frequency.

RAND, for its part, faces a problem of its own with respect to independence. The struggle for independence is not something that is ever definitively "won." It is a continuous process of adjusting interests, bargaining, sometimes compromising, sometimes holding firm in the myriad day-to-day decisions that have to be made in interaction with the client. Indeed, new pressures have arisen to threaten RAND's independence as never before. But if the quest for independence is difficult and uncertain, it is also eminently worth the struggle. For "the principal value of independent organizations is, plainly, their independence."[37]

Thus far the question of independence has been viewed mainly from the advisory organization's perspective. What of the sponsoring agency's requirements? A persuasive case can also be made here for allowing the advisory organization considerable independence. First of all, a pragmatic test of results seems to suggest that the agency will get more for its money if it imposes a minimum of administrative control on the research operations. Second, accountability in policy

[37] Richard Rovere, "Letter from Washington," *New Yorker* 36:120 (February 27, 1960). Rovere, writing in 1960, saw the "atmosphere" in which the advisory organizations were operating as one of growing political pressures and constraints. He went on to warn: "If restraints of the kind now imposed on the research organizations holding contracts with the Military Establishment are widely imposed and accepted, then the Government will lose the services of some gifted authorities and public opinion will be impoverished by the loss of many voices that might enrich it."

formation requires that the advisory organization be kept sufficiently separate and independent from public officials to avoid any confusion of roles or blurring of responsibility for decisions taken. In this context it may be appropriate to recall the Bell Report's warning that "in recent years there have been instances—particularly in the Department of Defense—where we have come dangerously close to permitting contract employees to exercise functions which belong with top Government management officials."[38]

The need for clearly separating the advisory from the government personnel thus suggests the desirability of a third kind of independence for the advisory organization— "physical (or geographical) independence." In addition to program independence and latitude on publication matters, it seems important to maintain simple physical separation of the advisory from the government personnel. This serves the contractor's interests insofar as physical independence contributes to program independence, and also serves the client's interests by preserving the unity of command and keeping a firm policy control in the hands of accountable officials. Where advisors are working in the same office with government officials (civilian or military), one can anticipate that ambiguities in status and division of responsibility are likely to arise sooner or later. Hierarchical separation is not always easily maintained under conditions of physical proximity; and mingling of contractor personnel in the administrative chain of command is an irregular practice that should normally be avoided.

A striking instance of a confusion of roles occurred in the early history of the Advanced Research Projects Agency (ARPA). In the 1957-1958 period, the Institute for Defense Analyses (IDA) provided technical-staff support for the

[38] Bell Report, in Holifield Subcommittee Hearings, part 1, p. 215.

agency which was "improper" for several reasons. In essence, IDA provided the major share of the agency's initial working staff. ARPA for some time had only a skeleton civil service staff. There was a blurring of the lines separating contract and government personnel.[39] The strange situation was also presented of contract personnel working alongside government employees and receiving higher salaries for essentially the same work, thus leading to a generally unhealthy working atmosphere. Further, research personnel were diverted into short-run administrative tasks for which they were not well-suited. Significantly, the solution to this situation was sought in the advisory corporation moving toward greater independence from the agency.[40]

Several further concrete instances may serve to illustrate the independence problem in its full intricacy. The first of these is the situation which culminated in the breakup of the Army's ties with the former Operations Research Office (ORO). The ORO, established in 1948, was the Army's first advisory organization (and has remained the Army's principal contract advisor). During the late 1950's, growing friction developed between Army officials and ORO Director Dr. Ellis Johnson.[41] Dr. Johnson, who had been director since the organization's inception, was a scholar of repute in the

[39] Based on confidential interviews with ARPA staff members.

[40] See Holifield Subcommittee Hearings, part 2, pp. 619-620. Mr. Richard Bissell, then President of IDA, made the following comment that is instructive in this context: "I can only say, especially over the last several years, especially as the relations of the IDA with the Defense Department have evolved, we have made every possible effort, with the cooperation of that Department, to cause them to evolve in the direction of greater independence.

This committee in previous years, I think, has been concerned with the early days of the Advanced Research Projects Agency.

I can say that today we are exceedingly careful to insure that, with minor and brief exceptions, no staff members of IDA serve any Government office in a capacity which could permit confusion between their status as non-Government advisors and the status of a Government officer."

[41] This reconstruction of events is based on confidential interviews. Any interpretations drawn from the events are my own.

field of operations research and a forceful man of strong convictions. Part of the difficulties stemmed from a clash of personalities, but also a root source of conflict was a difference in outlook between Johnson and Army officials about what independence meant for the advisory organization. Johnson, for example, held the view that the client should not interfere in the formative stages of the research. Numerous Army officials, on the other hand, believed that the Army needed a stronger voice in determining the direction of the ORO research effort. Another point of contention involved publication policy where sharp differences of opinion arose from time to time over whether material required a security classification or could be published in the open literature.[42]

The pressures came to a head in the spring of 1961. The Army decided to sever the ORO contract with Johns Hopkins University and to reestablish the ORO as a new nonprofit research corporation: the Research Analysis Corporation (RAC).[43] In the shuffle, Johnson was forced to resign and a new director was subsequently named to head RAC. The new organization was essentially the old ORO with a new director. Dr. Johnson issued a parting blast at Army officials for applying restrictive controls over the publications of ORO studies and for trying "to run the research—and that we couldn't tolerate."[44] Another senior ORO staff member reportedly observed that the new initials stood for Relax And Cooperate. For their part, Army officials complained of security breaches under Johnson's directorship and of numerous instances of failure to respond to legitimate Army needs. In a way, however, the

[42] In one dramatic incident, Dr. Johnson openly challenged the Army to arrest him for an alleged security breach.

[43] *New York Times*, May 28, 1961, p. 1, col. 4. The new contract with RAC became effective on September 1, 1961.

[44] Quoted in *Missiles and Rockets*, June 5, 1961, p. 11.

Army paid a backhanded tribute to Johnson by preserving virtually intact the organization that he had built.

A second illustration of the independence problem concerns an ambiguous relationship that existed between the Institute for Defense Analyses (IDA) and the in-house Weapons Systems Evaluation Group (WSEG). The main difficulties came clearly to light in 1962 when IDA and the Department of Defense sought to clarify the relationship between IDA and WSEG. IDA, it will be recalled, was formed in 1956 to provide technical support for WSEG. The latter group at the time was, by general agreement, not performing its responsibilities effectively. Accordingly, a Weapons Systems Evaluation Division (WSED) of IDA was given the job of assisting WSEG's civil service and military personnel in carrying out research projects. WSEG was to levy tasks upon WSED, arrange for access to information, provide military officers to assist in the tasks, and through the device of a review board to review finished work and arrange for its distribution and publication. There were, however, uncertainties in the assignment of responsibility between the two groups that proved troublesome. One individual, for example, served in a dual capacity as Director of WSED and as Director of Research for WSEG. "This arrangement," as former IDA President Richard Bissell wrote to Congressman Holifield in October 1962, "created a real ambiguity as to the responsibility for work performed by that part of the institute's research staff . . . under his [the WSED and WSEG Research Director's] supervision. It was not clear . . . whether this company as a Government contractor or, alternatively, the Director of the Weapons Systems Evaluation Group was ultimately responsible for the finished work."[45]

[45] Letter of October 1, 1962, reprinted in Holifield Subcommittee Hearings, part 2, pp. 633-635.

273

Moreover, Bissell went on to note, "a legitimate doubt could arise as to whether this ambiguous relationship did promote 'the detached quality and objectivity' of the work performed by the contractor, which was asserted by the Bell Report to be one of the principal advantages which the Government might hope to realize from the subcontracting of research to private organizations."[46]

IDA management thus became convinced of the need for some changes in the IDA-WSEG relationship. In the spring of 1962, after the Bell Report's release, IDA management entered into serious discussions with senior Defense Department officials with that purpose in mind. The initial result of these discussions was a memorandum of 11 July 1962 from Dr. Harold Brown, Director of Defense Research and Engineering, to Lt. General William P. Ennis, WSEG Director, ordering certain changes in the IDA-WSEG relationship.[47] In essence, the changes attempted to make IDA's WSED more independent from WSEG and from close supervision by the service representatives attached to WSEG. The main features were discontinuance of the practice of having an IDA official serve in a dual capacity as Director of Research for WSEG, adoption of the requirement that IDA submit its studies directly to the Joint Chiefs of Staff and to the Office of the Secretary of Defense instead of to WSEG for review, and reorienting the WSEG Review Board from its position in the chain of operations to a purely advisory position.

The changes evoked a storm of protest from parts of the professional military.[48] Military officers objected that the

46 *Ibid.*, p. 634.
47 The memorandum is reprinted in full in Holifield Subcommittee Hearings, part 2, pp. 636-637.
48 See Hanson Baldwin, "Pentagon Edict Upsets Military: Officers Fear Curb on Role in Weapons Evaluation," *New York Times*, July 28, 1962, p. 2, col. 1.

changes would effectively eliminate any military control over WSEG's studies, operations, and reports. Consequently, the group's work would be less responsive to actual military needs. There was also some concern about security requirements which many military men felt would be relaxed to a dangerous extent under some of the projected changes. The overriding worry in the military, however, was that the changes would reduce still further the influence of professional military judgment in the defense decision-making process.[49]

In a strongly worded memorandum to the Director of Defense Research and Engineering, General Ennis attacked the announced administrative reforms as undermining the basis for effective civilian scientist-military cooperation in WSEG. For all practical purposes, he charged, the changes would render the positions of WSEG Director and the senior military representatives superfluous.[50] General Ennis also asserted that the changes placed "the military and civil service personnel . . . in a position of being free labor for a civilian contractor, with the resulting further downgrading of the military and civil service personnel of DOD."[51] One can sympathize with this reaction. Although the Brown memorandum attempted to clarify the dual role of the IDA official who served in the WSEG chain of command, it did so by vesting clear authority with the contractor. In effect, the government personnel were left with a highly ambiguous and minor supporting role to the contractor. If contractor personnel should not be closely supervised by government em-

[49] Hanson Baldwin quotes a senior officer as having told a Pentagon civilian official, "I think I'm going to let you have my uniform and you can fight the next war with the weapons you design." *Ibid.*

[50] Memorandum of July 16, 1962, reprinted in Holifield Subcommittee Hearings, part 2, pp. 635-636.

[51] *Ibid.*, p. 636.

275

ployees, it would certainly also seem important that government employees should not be supervised by contractor personnel.

Faced with a strong military counterattack, the top Pentagon civilian officials were forced to reconsider their decision. A Department of Defense Instruction was subsequently issued on 23 August 1962 which sought to clarify WSEG's role and which represented a partial backdown from the earlier Brown Memorandum's objective of greater independence for IDA.[52] A distinction was to be made between a WSEG and a contractor study, each to be carried out somewhat independently but to be mutually supporting. The WSEG Director retained the authority to "assign tasks and projects and their priorities to contractors and will receive and forward all reports and communications relative to these tasks and projects."[53] In general, the new instruction seemed at best a stop-gap arrangement that provided only a tentative and uneasy solution. Many ambiguities and uncertainties remained to be worked out in practical working relationships under the new arrangement. It appeared likely that further organizational changes might follow. One change already in prospect was IDA management's decision to move WSED physically out of the Pentagon to an adjacent building. Although WSED would continue to work closely with WSEG, the change would afford some measure of physical independence for the contractor and would presumably help avoid some misunderstandings and role confusion. This change might help point the way toward a more stable and effective long-term relationship between IDA and WSEG.

At this point, however, it is difficult to draw unambiguous

[52] Department of Defense Instruction 5129.37, August 23, 1962, reprinted, *ibid.*, pp. 637-639.
[53] *Ibid.*, sec. VII, para. 13, p. 638.

"lessons" from either the RAC or the IDA-WSEG experience. But these cases provide graphic illustrations of the continuing intricate problem of assuring the advisory organization sufficient independence to perform useful research yet tempering that independence to keep the organization responsive to client needs.

THE COMMUNICATION OF RESEARCH RESULTS

The problem of effectively communicating research and advice to policy makers is related to the question of the advisory organization's independence. For the nature and extent of the independence clearly has some bearing on the advisory organization's ability to gain the ear and the confidence of government officials. The communication problem, however, goes beyond that of independence and must be considered in a context of its own. The independence factor can provide the preconditions for, or set the outer limits to, effective communication; but it will not itself guarantee or preclude effective communication. All of the defense "think" organizations, in one form or another, must concern themselves with this problem. To be sure, those organizations which have intimate working contacts with client personnel will have a different sort of communication problem than the organization like RAND with considerably greater independence and almost no direct client participation in the research effort. Our remarks here will focus chiefly on the problem as viewed from the RAND perspective.

A first general "lesson" to be learned from the RAND experience is that effective communication of policy research begins *before*, not *after*, the publication of a study or the preparation of a finished report. This is a lesson that RAND people have learned the hard way—through contemptuous responses and sometimes plain indifference to studies that have taken months and even years of hard work. Effective communication generally requires that appropriate client officials (and this may include the operating levels as well as the higher policy echelons) be apprised of the study's existence while the research is still in progress. Ideally, the client's interest and enthusiasm should be elicited at some point and a sense of involvement in the study's progress should be created through occasional informal consultations. This helps assure a careful hearing for the study and whatever policy recommendations it may contain. And the disturbing impression is avoided, from the policy maker's point of view, of being confronted with a document prepared in the abstract without an understanding of the agency's "real" problems.

The communication of policy research, in brief, presents a different situation in many respects from the scholar wishing to communicate the results of basic research to a society of academic peers. Publication alone cannot be relied upon as the major means of communication.[54] The audience for whom the research is intended is not a group of academic colleagues with the leisure or inclination to spend considerable amounts of time digesting the contents of lengthy reports. The audience, rather, is a group of harried officials whose attention is constantly distracted by numerous pressing concerns. A big problem often may be simply to gain the attention of officials. A premium also must be put on ways to make information available rapidly and at timely points in the decision-making stream.

The organization like RAND thus must resort to a number of supplements to (and sometimes substitutes for) the prac-

---

[54] Even in academic research the adequacy of normal publication channels has been questioned. Normal publication channels have been criticized for being too "slow," and for failing to keep abreast of the latest developments at the frontiers of knowledge. See Gerald Holton, "Scientific Research and Scholarship: Notes Toward the Design of Proper Scales," *Daedalus*, Spring 1962, pp. 362-399.

tice of open publication and scholarly exchanges that form the normal pattern for disseminating research results. One of these—the technique of the "briefing"—was mentioned above in our discussion of the basing study. The results of the basing study were thoroughly briefed, discussed, and disseminated through the Air Force before a final version of the study had even appeared. The circulation of a draft report is frequently another valuable way to communicate research results and to arrange for a "feedback" of client views that might prove useful in the final version of a study or in later research. Another practice that RAND has used less frequently consists in inviting small groups of officials from the sponsoring agency out to RAND headquarters to discuss the research informally prior to writing a draft. Or researchers might travel to Washington for discussions at various stages in a study's preparation. Beyond this, there is a wide range of formal and informal tactics that may prove useful in the communication process. The employment of any specific tactic must be a matter of judgment and discretion in a particular situation.

Aside from any formal techniques, however, simple human tact and courtesy are obviously important factors in effective communication of policy research. Traditionally, RAND has faced something of a *prima donna* problem, particularly in its relations with the military. As Bernard Brodie remarked in a masterpiece of scholarly understatement: ". . . it has always been true that creative abilities are not necessarily combined in the same person with such character endowments as tact and modesty."[55] Client egos have frequently been bruised by the researcher who has little patience with the

---

[55] "The Scientific Strategists," *Administration of National Security: Selected Papers*, Committee Print, Subcommittee on National Security Staffing and Operations, Committee on Government Operations, U.S. Senate, persuant to S. Res. 332, 87th Cong., 2nd Sess., p. 200.

official's failure to grasp a point in a report or briefing. This is, of course, a continuing problem of human relationships for which there is no ready cure. Yet the advisory organization can and should encourage tactful deportment on the staff's part in dealings with the client.

Granted that communication should not wait until a study is completed, is there any special point at which the communication process should normally begin? Some observers conclude that "the process of communication of the results of operational research should start at the beginning of a project."[56] Under this view, concern with how, and to whom, the research will be communicated should be integral parts of the researcher's job from the very beginning, along with close cooperation with the client in framing the issue to be investigated. RAND has in general rejected this alternative in favor of beginning the communication process at a later point— usually after some significant preliminary results have been obtained. Although the relatively late stage at which the process is initiated does complicate the task to some extent, there seem to be compelling reasons to err on the side of tardy rather than premature communication. The research project requiring client consultation in its formative stages may be hampered at subsequent stages by an overly restrictive definition of the problem to be studied.[57] The numerous dangers of

[56] Pamphlet, B. D. Hankin, "The Communication of the Results of Operational Research to the Makers of Policy," reproduced with a foreword by Ellis A. Johnson, Operations Research Office, Johns Hopkins University (originally published in *Operational Research Quarterly*, July 1959), p. 10. See also Russell L. Ackoff, *Scientific Method: Optimizing Applied Research Decisions* (John Wiley, New York, 1962), ch. xiv. Ackoff notes (pp. 408-409): "Planning for the implementation of research results should begin when the research itself begins; it should not wait until the results are obtained. Specifically, the technical abilities of those who will use the results and the facilities at their disposal should be taken into account in determining the form and nature of the research results which should be sought."

[57] Witness RAND's experience with the SOFS project. See Ch. V above.

research that is too closely controlled by short-run client interests may apply, and especially the danger that merely "symbolic" (that is, innocuous) research will result.

Further, research that is carried out with the objective from the start of "influencing" certain carefully chosen individuals may acquire an air of bias and disingenuousness. An element of subjectivity enters that is not easily reconcilable with the standards of disinterestedness that normally govern good research work. Whether or not one believes that research can ever be truly "objective"—particularly research in the inexact sciences—it would be difficult to maintain that the researcher should consciously abandon the quest for objectivity and should instead engage in artful equivocation and cautious deception in the interests of increasing the research product's "saleability." In the end, such a tactic would be damaging for the researcher and self-defeating for the research organization.

There is probably no precise answer to the question of when the communication process should begin. But the following seems a useful approximation: it should normally begin as soon as the researcher feels that he has obtained some significant preliminary results in his work. The initial efforts at communication will usually be in a low key, taking advantage of preliminary client reactions to check the realism of assumptions, and finally will intensify as the study nears completion. For the policy study whose primary aim is to force a rethinking of some problem—and which has no clear "results" or specific recommendations—it is still important to devote effort to communication so that the research is known, understood, and fully considered in official thinking.

The timing, manner, form, and extent of the communication will depend upon specific circumstances and the preferences of the researcher. The term "researcher" can be taken almost literally here. The individual researcher is generally the best judge of whether he can spend his time more fruitfully in further research or in communication efforts. The individual researcher should also have a large voice in deciding when a study is ready to be communicated, though there will normally be some review process for external publications.

The communication process should continue until there has been a wide exposure of the research throughout all affected parts of the client agency. In some cases, this can mean that the communication process should continue even after the study has influenced important decisions. For the "operating" as well as the higher policy levels must understand the significance of a piece of research if effective implementation is desired.

A last general "lesson" from the RAND experience in this sphere may be inferred from what has already been said. It is that, in the final analysis, the communication process is not primarily a matter of discovering better techniques of presentation or ways of packaging information in an attractive fashion. Rather, what is principally at issue is the substantive content of the research or advice to be communicated. The crucial aspects of the process center around the worth of the research and (in the case of a briefing or other oral presentation) the researcher's grasp of his material. The research will be successfully communicated if it can survive the critical scrutiny of the user officials (or even if it appears solid enough merely to invite the critical attention of busy officials). It follows that it is very important to have the knowledgeable researchers who have actually done the research carry the major burden of communication, and to guarantee that only high quality work is communicated to the client.

In closing, it may be well to recall B. D. Hankin's challenge to the analyst: "In operational research . . . to publish . . . is not nearly enough. In the broadest possible sense of the word we have to *communicate*. . . . The aim must be to get the results and conclusions studied, respected, understood and fully considered in the formulation of policy."[58]

THE STAFFING PROBLEM

A chief remaining topic for consideration is the necessity for RAND to assure a continued supply of qualified researchers. As previously noted, RAND benefited greatly in the early years from the fact that large numbers of researchers who had served in the war effort were in the process of returning to civilian life and were available for employment. The disruption in the normal career patterns meant that intellectual skills could be recruited in much larger numbers than would normally be the case in a more settled period of career development. More important, some of the outstanding individuals who did so much to build RAND's reputation for distinction were obtained at this time.[59] Another (somewhat less important) infusion of skills occurred during and immediately after the Korean War. But by the beginning of the 1960's the staffing problem had gradually grown more difficult. There was a normal attrition as the original leaders had either departed or had passed their creative peaks. Changes

[58] Hankin, "The Communication of the Results of Operational Research to the Makers of Policy," p. 1.

[59] The case of the late John D. Williams is a notable example. Williams was a brilliant mathematics graduate student at Princeton at the time World War II broke out. In part through the influence of John von Neumann, whom he greatly admired, Williams was drawn into service in the war effort. At the close of hostilities, Williams was considering his return to the academic world to complete his graduate studies when the invitation came to join the novel Project RAND undertaking. He accepted what seemed to be a challenging assignment, and his presence at RAND facilitated the recruitment of other top mathematicians. RAND's success in large part rests on the good fortune of having men of John Williams' stature as early additions to the staff.

in RAND's external environment and internal administration also began to have an effect on staffing. Increasingly, newer skills were also needed (particularly social scientists of various kinds) to assume the roles of "generalist" and "integrater" in the policy studies. But RAND had to compete for these skills in an increasingly competitive market. The net result of all these factors was that RAND began to face a serious problem in maintaining the quality of the research staff.

In a broad sense, the staffing problem is bound up with a complex nexus of issues concerning RAND's future and cannot sensibly be considered in isolation from them. Serious interference with RAND's "independence," for example, would surely have a profound impact on the staffing question. However, for the moment, let us attempt to hold these broad factors more or less constant, and focus our attention instead on several more immediate aspects of the staffing problem. RAND's competitive positions with respect to the government service and to the academic community require special attention. For the government service and the universities are chief competitors for an important range of talent in which RAND is interested. This range of talent consists chiefly of research administrators of various kinds and a class of what might be very roughly termed "activist academics." There will always be talented individuals who prefer government service because of the sense of direct involvement it provides. There will be another temperament that only feels at home in a university setting. But in between there is a group of individuals with the requisite training and skills who might be swayed to either government service, the academic community, or employment in the RAND-type institution for all or some part of their careers. It is these individuals for whom the organization like RAND is principally competing. It may be assumed that these individuals are sensitive to various advantages, including salary, offered by each type of employment (and each employer).

RAND also obtains a certain range of skills from industry, but does not in the usual sense "compete" with industry directly for talent. RAND will enter the picture usually only after an individual has decided to forsake the generally higher industrial salaries for employment with other compensations, and here again the government and the universities are major competitors (along with certain of the sister advisory institutions).

Let us first consider RAND's position with respect to the career service. Here RAND faces a paradoxical situation. On the one hand, there is a natural desire to attract a fair share of the top talent. Yet, at the same time, reasonable prudence would suggest that the organization like RAND should not want to do unusually well over long periods for fear of possible grave political repercussions. The Bell Report has done much to focus attention on the issue of the inadequate salaries and working conditions of scientists and research administrators in the career service.[60] A serious and persistent disparity in compensation for comparable training and skill levels between the advisory corporations and government agencies could threaten the advisory corporation's chances for survival in the American governing system. Organized opposition from the federal career service might force the adoption of legislation to curtail drastically the advisory organization's autonomy in regard to salary and personnel practices. Already certain civil service and congressional groups seem to have formed a tentative alliance against the contract advisors. Their aim seems to be to protect the interests of the career

[60] The Bell Report, in Holifield Subcommittee Hearings, part 1, esp. 216-218 and annex 5, pp. 279-304.

servants and to safeguard Congress' voice in policy formation. Presumably career servants are, in the view of many congressmen, more accountable to Congress than are contract employees. Thus it would seem important for the organization like RAND to reach some kind of accommodation with the government career service.

Indeed, the need for achieving some kind of *modus operandi* with the career service is not merely a matter of narrow political expediency. In the long run, the concept of the semiprivate organization functioning as research arm and privy counselor to the government only makes sense if some kind of approximate balance exists between the government's and the outside group's attractiveness as an employer. Unless the government is able to attract some minimum share of the individuals with the appropriate training and ability in the research field, the long-term effect will be an erosion of the government's capacity to evaluate the advice it receives. In a word, the whole political center of gravity could be shifted and the government agencies could become the satellites of the private institutions, rather than vice versa.[61] Such a development would be sharply at odds with our notions of democratic responsibility. And it would, in the end, present problems soluble only converting the contract employees into a new bureaucracy with clear channels of accountability to the President and to Congress.

The organization like RAND, therefore, must steer a difficult course with apparently contradictory objectives as guides. It must, on the one hand, pay the salaries and provide the sort of working environment that will make employment opportunities in some ways appear more attractive than gov-

[61] For an elaboration of this point, see Don K. Price, "The Secretary and Our Unwritten Constitution," in Don K. Price, ed., *The Secretary of State* (American Assembly, Prentice-Hall, Englewood Cliffs, N.J., 1960), pp. 169-170.

ernment service. But carried to its logical conclusion this approach is self-defeating. Hence the advisory organization cannot be too obviously and consistently successful in recruitment, especially in attracting the individuals who might otherwise be interested in government service. Certain steps should be taken to show good faith or, put negatively, actions which give unnecessary offense should be avoided. For example, one can imagine that the advisory corporation which makes a practice of "raiding" government agencies and hiring away top talent would have some difficulties in working harmoniously with clients. For its part, RAND does a number of things to help maintain an equilibrium with the career service. RAND makes it a practice, for instance, not to hire people away from the government at a higher salary than they were receiving in government employment. This "no raise" policy has been in effect at least since the late 1950's. RAND also will not normally consider hiring any government employee who has not already for independent reasons decided to leave government service.[62]

In actuality, it does not seem impossible to reach some kind of equilibrium in the relationship between the contract advisors and the career service. Government salaries for scientists, social scientists, engineers, and research administrators have made some substantial gains in recent years, although it is possible that RAND may still be slightly ahead in certain categories.[63] The career service can be expected to strengthen

[62] Statement of RAND President F. R. Collbohm, in Holifield Subcommittee Hearings, part 3, p. 928.

[63] Career service gains are reported in John W. Macy, Jr., "The Scientist in the Federal Service," *Science* 148:51-54 (April 2, 1965). RAND may be slightly ahead with respect to salaries paid to senior researchers and to new Ph.D.'s, two categories RAND seems to consider particularly important. It must be stated that my discussion of salaries rests on what appears to be fairly common knowledge "in the trade." The discussion is therefore somewhat conjectural. I have had no access to specific salary data of the RAND Corporation,

its position further with narrowing salary differentials between government and contract employment and with continued improvement in working conditions within the government. The widespread recognition of the need suggests that the government will increasingly obtain adequate numbers and quality of skills to provide a substantial scientific and technical capability. Progress toward this end will be hastened if the contract research community shows restraint and does not attempt to negate the effect of government salary improvements through offsetting salary raises.[64] The contract advisors should certainly be encouraged to show such restraint, and to recognize that their continued existence as independent organizations depends on a strong and competent government service.

It is perhaps doubtful, however, whether RAND could follow precisely the practice of Analytic Services, Inc. (ANSER). (ANSER sought to avoid problems with the civil service by, in effect, adopting civil service policies with respect to salary, promotions, wage increases, and the like.) RAND is bidding for a different range of talent, and its

---

or any other research organization. In the trade, however, the Institute for Defense Analyses (IDA) enjoys a reputation for paying the highest salaries in the family of nonprofit advisory organizations, with RAND second, and the rest of the nonprofit advisors dropping off rather sharply. The systems management corporations like Aerospace and MITRE appear to pay salaries substantially above the advisory corporations. A possible explanation for this is that they must compete directly with industry for certain ranges of talent. For a general description of the RAND salary structure, see *Department of Defense Appropriations for 1963*, Hearings before a Subcommittee of the Committee on Appropriations, U.S. House of Representatives, part III, Operations and Maintenance, p. 678.

[64] In the Holifield Subcommittee Hearings, some fears were expressed that contractors could negate the effect of a government salary raise simply by following with salary raises or fringe benefit improvements of their own. Conceivably, the artful contractor could strive to conceal such maneuvers by juggling around the contract cost figures. See the discussion in the Hearings, part 1, pp. 18-24.

flexibility in personnel policies is an important asset in obtaining individuals who might not otherwise be available for work on public problems. It is perhaps desirable for the government to move toward similar flexibility in personnel practices, but in the meantime it would make little sense to degrade one capability without materially improving the other. The principal objective is not to constrain the contractor but to revitalize the government.

Further, a certain fluidity in career patterns suggests that there are some direct mutual interests in the civil service's relationship with the contract advisors. If the flow of talent were all one way, or if an initial career choice tended to lock one into a rigid pattern, the situation would be much more difficult. In fact, there seems to be a considerable movement of people from the advisory corporations to the government service, and vice versa. Although empirical evidence is scant, the flow does not seem marked by a great imbalance in either direction.[65] Indeed, RAND's role as something of a training center for certain talents that eventually may be channeled into government service should contribute to a strengthened career service in the defense agencies.

In brief, the normal pull of market incentives which upgrade the attraction of government service, combined with self-restraint shown by the contract-research community, seem

[65] In the period from January 1, 1959 to August 1, 1965, for example, RAND lost 33 professionals to the government service, while gaining 24 individuals from the civilian career service and 12 former military officers. Material gathered from the RAND personnel office. The numbers alone, however, may be somewhat deceptive. An important fact to be kept in mind is that some of the individuals who have left RAND to join the government have been senior people of high quality, and thus the government may have benefited considerably from the exchange. The figures also do not include the individuals who eventually joined the government service after working in RAND's Bethesda office in connection with the assistance RAND rendered the Defense Comptroller in establishing a program budget system. For a brief description of this activity, see note 12 of Chapter V above.

to offer hope that an accommodation can be reached without essentially altering the pluralist institutional base that has evolved in America's government-science partnership. The consequences of more drastic action appear rather disturbing. To attempt some kind of direct control over contractor salaries and related benefits seems to pose formidable administrative problems and to imply great increases in government power.

Consider, for example, the Bell Report's suggestion that all contractor salaries above a certain level—say $25,000—should require the personal approval of an official reporting directly to the agency or department head.[66] The intent of this recommendation was, perhaps consciously, left somewhat ambiguous. It was uncertain whether the report envisaged a strict upper limit to be exceeded only in exceptional cases or whether the approval would be rather perfunctory in nature. In any case, considerable alarm and confusion spread throughout the contractor community when some industry spokesmen expressed their belief that the Kennedy Administration intended to place a "top limit" of $25,000 on all salaries in research and development organizations awarded contracts in noncompetitive bidding.[67] Sufficient apprehension was generated to cause David E. Bell, then Budget Bureau director, to attempt to clarify the administration's position in a letter to the editor of the trade journal *Missiles and Rockets*. Bell denied that the Kennedy Administration contemplated any top limit on salaries, and declared that it was explicitly assumed that the government should continue to reimburse for salaries above $25,000 "in appropriate cases."[68] What constitutes an "appropriate" case, however, has remained obscure.

[66] The Bell Report, in Holifield Subcommittee Hearings, part 1, p. 240.

[67] See the editorial, "Curiouser and Curiouser," in *Missiles and Rockets*, July 9, 1962.

[68] Letter to the editor, *Missiles and Rockets*, August 20, 1962.

It is difficult to determine how effective this salary control would be if such a proposal were actually implemented. The numbers of people involved who received salaries above $25,000 would be an important factor. Current estimates, based on partial surveys, range from a figure in the several hundreds to one in the several thousands or even higher.[69] If the number ranged into the thousands, the task of administering the review procedure would pose formidable problems. The government would have to gather accurate and up-to-date information, create new machinery, and attempt a difficult wage arbitration task for parts of the economy. There must be some point at which the "administrative costs" of an action exceed the expected benefits. Moreover, the government would be likely thrust on an unprecedented scale into the internal life of private organizations. Internal management responsibilities such as hiring, promotions, and awarding special pay incentives presumably would be subject to some ill-defined public scrutiny. One hopes that such controls will not be necessary. A policy of self-restraint in salary matters on the part of research contractors, avoidance of "pirating" talent from government agencies, and serving as a training center for channeling some individuals into government work hopefully can do much to normalize relationships with the civil service. Meanwhile government service opportunities will be growing more attractive. Thus an approximate balance in the attractiveness of government and contractor employment may be attained—and without the difficult and disturbing prospect of some form of wage control over sectors of the economy.

The reverse side of the coin with respect to the staffing problem is RAND's relationship to the academic community. Here again RAND finds itself in a paradoxical situation. In-

[69] See the Holifield Subcommittee Hearings, part 1, pp. 32-34.

creased academic interest in defense studies in a sense has served RAND's ends by widening the reserves of relevant talent, removing traditional prejudices against the study of military affairs as a legitimate field of academic inquiry, and promoting a dialogue on defense problems among intellectual elites. But in another sense it has served to downgrade RAND's relative importance and to decrease somewhat its attraction as a major research center. Individuals could now pursue research interests in the national security area within an academic setting. Further, the rise in academic salaries and the growth of consulting opportunities in recent years have greatly reduced the financial incentives for leaving the academic world to join the institution like RAND. As academic salaries have increased, RAND has sought to stay ahead in important brackets (senior researchers, new Ph.D.'s). But the political sensitivity of the issue of nonprofit corporation salaries has hampered this effort to some extent. A slightly fanciful analogy might depict RAND in a situation akin to the supposed French desire in the nineteenth century for a German army strong enough to defeat the Russians but not strong enough to cross the Rhine. The organization like RAND would like to keep salaries slightly above academic salaries but not so far above that a dangerous political issue is activated with Congress and the civil service. In any case, there can be little doubt that growing academic competition has added substantially to RAND's staffing problem.

In the face of growing difficulties in attracting qualified researchers away from the universities, the organization like RAND has a number of alternatives. One cluster of possibilities that comes to mind is to forge various closer ties with the academic community. The advisory organization, for example, might seek to locate only in the vicinity of important university centers where talent could be "shared" with university departments.[70] To a limited extent, this tactic has been used in RAND's case. A few appointments have been made on a "joint" basis with a department of the University of California, and at other times individuals have come to RAND with an understanding that they can teach occasional courses at local (or other) institutions. This has enabled RAND to attract some people who wanted to maintain an academic tie and who needed the stimulus of university teaching. One can imagine a number of other possible closer ties with the academic community that could be important for the organization like RAND. There are, however, some serious drawbacks to these suggestions. A question of institutional "identity" would surely arise if sizable parts of the staff had divided loyalties and were not clearly subject to an over-all management supervision. One can imagine that a research organization's management and a university administration would both be unhappy with any arrangement that created ambiguities in the status of substantial numbers of employees. Some resentments would probably also be generated within the research organization's staff between those who had the perquisites of a university affiliation and those who did not.

RAND could also develop something like the academic tenure system as a possible means to attract senior people from the universities. This would, however, negate one of

[70] Alvin M. Weinberg, director of the Oak Ridge National Laboratory, has made a number of interesting suggestions along these lines. See his "The Federal Laboratories and Science Education," *Science* 136 (April 6, 1962). Weinberg sees the possibility of the great national laboratories assuming some kind of role in graduate education in the applied sciences, possibly even conferring advanced degrees (as, for example, the Brookings Institution did until the late 1930's). Difficult questions of institutional specialization are involved here, but this sort of suggestion has some intuitive appeal and deserves further careful thought. A major challenge might be to consider ways to effect a "co-location" of research institutions and university centers so as to promote a greater geographic dispersion of scientific and intellectual talent and to avoid a further aggregation of talent in areas that are already intellectually "rich."

293

RAND's attractions. A number of people come to RAND in part to escape traditional academic status distinctions.

Another alternative is simply to rely increasingly on the device of the consultant to obtain needed skills. This has some advantages, for skills can be obtained from a broad geographical and institutional base. Often top talent may be available for temporary periods that could never be obtained otherwise. Extensive consulting contacts with the academic community have always been a standard feature of RAND's method of operation, and presumably it would not be very difficult to augment substantially these contacts. But this course has disadvantages similar to those outlined above. Effective management of the research effort could become difficult if more and more work were parceled out to consultants whose major professional interests lay outside the organization. Moreover, temporary consulting arrangements seldom exploit fully what is RAND's special strength—the interaction of different professional skills to achieve results beyond those of a single discipline. Consulting assignments done outside the organization by definition do not have the opportunity for interaction with other RAND research skills. Even the consultant who serves in residence at RAND, however, seldom benefits fully from the interdisciplinary environment because his stay is usually relatively brief. The organization like RAND doubtless gains a great deal by having a number of academic consultants, and should strive to keep a reservoir of talent available for various needs. But it is illusory to expect that a roster of consultants can serve as the nucleus of a staff.

Still another possibility for RAND is to devalue the currency—to settle for a range of talent for which there is not such keen competition. This, for reasons suggested below, is perhaps the worst choice that RAND could make. It seems preferable to maintain the quality of the research staff even if this means little or no expansion in staff or in the responsibilities that the organization can assume. Fewer research tasks performed well would seem to mean much more to RAND than a larger number of lower quality efforts.

We must now reintroduce those general considerations that have been kept in the background of the discussion. For in the end these broad factors may prove decisive. If RAND can continue to enjoy a considerable measure of independence, work on problems of central national importance, and avoid a bureaucratic internal administration, the chances are good that it can maintain a high-quality staff even in the face of increased competition for talent. So long as there are exciting research opportunities, it should still be possible for RAND to attract a share of the talent. If there were a loss of independence, less interesting research foci, and bureaucratization, one can anticipate a decline in the general quality of RAND's staff. With it there would surely be a decline in RAND's significance. The organization possibly would be transformed gradually from a research into a "service" organization. The danger is not that RAND would cease to exist: its past achievements and relatively long history probably guarantee a continued existence. The danger is, rather, that it may grow sterile.

A matter of crucial importance for RAND, therefore, is to maintain consistently high standards in staffing. A "vicious circle" tends to result when compromises are made in staffing. More direction and more rules are required to administer a research organization when one is not quite sure of the staff. But the rise in visible administration may frighten away or drive off the better individuals. Thus management in turn must reach still lower. This brings new problems of internal administration, more need to oversee operations, and in turn results in a further deterioration in staff quality. The quality of the advisory organization's staff is also related to the problem of maintaining independence and in being asked to participate in the debate on significant issues. The nature of the research assignments and the degree of independence both affect, and are affected by, the quality of the research personnel. Individuals of stature are likely to command the respect of client officials and to help assure a high status for their organization.

► VIII ◄

CONCLUSION

During the past several decades, the United States has witnessed revolutionary changes in its position in the world and in the dynamics of its governing system. At the root of these developments are explosive advances in science and technology that have placed instruments of vast destructive power in the hands of rulers and enormously increased the complexity of the policy process. The advent of nuclear weapons has introduced lasting changes in the relations among nations. The United States has been forced into a position of world leadership, and the security hitherto afforded by geographic separation from unfriendly countries has become illusory. Domestically, scientific and technical elites have become important participants in the processes of policy formation. The traditional distinction between "public" and "private" has lost much of its former meaning as many private institutions carry on public functions through the device of the administrative contract. It has also become clear that the lines between civilian and military can no longer be as sharply drawn as they were in the past. The nuclear age has witnessed a growing military influence in areas once regarded as civilian. Conversely, there has been a certain "civilianizing" of the military as nonmilitary expertise has become essential to military planning and operations. The complex interplay between domestic and international affairs also has forced a revision in the old notion that "foreign" and "domestic" affairs were distinct and separable fields of activity and thought. In brief, the nuclear age has drawn together government and science, public and private institutions, civilian and military personnel, foreign and domestic concerns into a complex new pattern of interrelationship and reciprocal influence—a pattern that cannot be understood in terms of our traditional folklore about the conduct of public affairs.

One aspect of this whole broad picture that has attracted considerable interest is the role of the contract advisory organizations working for defense agencies. This study has sought to clarify that role. The objective has been to suggest some of the real strengths and weaknesses of this novel type of organization, to relate the advisory function to the policy process, and in general to raise the level of understanding of a new and little understood area.

We saw that the RAND-type advisory institution can serve a number of important functions in America's pluralist governing system. The organization like RAND helps to keep open channels of communication for debate and discussion of broad policy issues; and it guards against the dangers of a closed system with men in power cut off from the healthy effects of criticism and advice from different points. By serving as a link between the "insiders" and a broader intellectual public, this type of institutional arrangement substantially reduces the danger that the free play of policy debate will be choked off within a narrow self-contained elite.[1]

RAND staff members will be in a position to know some of the policy maker's major concerns and some of the crucial dimensions of important problems. And having the freedom

---

[1] See Chapter VI above.

to think full time about these problems, in a stimulating environment, they may occasionally generate useful new ideas and anticipate some of tomorrow's problems that a swiftly evolving technology may bring. In the first instance the results of their work will normally be for the consumption of government officials only. But later the research findings and methodologies usually are widely aired through unclassified writings and speeches to a broader intellectual audience. Through such unclassified writings and speeches, and through RAND's consultantship links with the academic community, RAND can contribute to forming a wider and more attentive intellectual public interested in national security issues. What is significant here is not that RAND has a kind of inside dopester's access to the specifics of important issues or that advisory personnel divulge sensitive information to unauthorized individuals. The important point is rather that the advisory corporation like RAND, close to yet somewhat aloof from the centers of power, can combine a sense of relevance in its work with the opportunity for posing new approaches to problems that may prove useful to the policy maker and the academic critic alike.

But on occasion a useful function may be served by simply providing reliable information for use in serious discussion. A good example of the contribution that the semiofficial group can make in this respect occurs even in Britain where there is a notably tighter control over information than in the United States and where there is no exact counterpart to the RAND-type advisory institutions. I refer to the annual estimate of Communist bloc and Western military strength prepared by the nongovernmental Institute for Strategic Studies in London.[2] The estimate contains force level figures, esti-

---

[2] *The Communist Bloc and the Western Alliances: The Military Balance,*

mates of the numbers of strategic and tactical weapons of various kinds possessed by the two blocs, expenditures on defense, and in general seeks to present an accurate picture of the two blocs' military strength. Originally, the annual estimate was prepared entirely from public sources and on the basis of shrewd inferences from information generally available. Subsequently, however, the British Defence Ministry, recognizing that the estimate figures were being widely used, has given unofficial guidance to the Institute in the annual estimate's preparation to assure that the figures are reasonably accurate. Hence the Institute's annual estimate has come to provide a fund of reliable information which scholars, journalists, and publicists throughout the West can use in debating national security issues.[3] Thus a great deal of obviously false information can be relegated to the penumbra of serious discussion. Again it should be stressed that this is not a case of "leaking" classified information. What is involved here is the fact that the semiofficial organization can afford to say things that a government agency for policy reasons cannot.

It is important to note, however, that RAND's role as a link between the "insiders" and a wider public is necessarily *and properly* circumscribed. It is not and should not be the role of the advisory corporation like RAND to deal directly with the general public in the exchange of information and ideas on broad defense issues. RAND's role rather is to deal with a

---

1962-1963, pamphlet (The Institute for Strategic Studies, London, November 1962). The following comments about the annual estimate are based on an interview with Alastair Buchan, Director, Institute for Strategic Studies, in London, January 3, 1962.

[3] For one of the few studies of the problems that may result when the "neutral" expert is drawn into political controversy and the accuracy of his information is challenged, see Kathryn S. Arnow, "The Attack on the Cost of Living Index," in Harold Stein, ed., *Public Administration and Policy Development*, The Inter-University Case Program (Harcourt, Brace, New York, 1952), pp. 775-853.

specific sector of the broader public—the intellectual community—and only in the most indirect sense to serve as communications broker between the government and informed citizens. It is the task of publicists, academic scholars, and others to complete the link and to educate the general public on problems of concern to RAND clients. The sort of role that RAND can play consists principally in encouraging debate, intellectual discussion, and further research on complex long-range problems. RAND is not and should not be concerned with public agitation for the adoption of particular solutions to pressing and controversial policy issues. Many fears have been expressed that RAND is somehow an instrument for influencing the views of informed citizens on controversial policy matters. To some extent, it is perhaps unavoidable for some spillover of RAND's intellectual activities to influence public attitudes from time to time. But this is distinctly not RAND's main function. More will be said on this point later. Here it suffices to point out that RAND can serve an important function in tapping the intellectual capabilities of people outside the government for work on public problems.

Further, the advisory corporation like RAND can play an important role in facilitating innovation in government organization. It is widely recognized that organizations, and especially large organizations, have great difficulty in effecting change in their policies and procedures. Research has begun to show how the internal dynamics of organizational behavior inhibit creative innovative activity; the sources of change seem often to lie outside the immediate organizational framework.[4] Thus the RAND-type advisory institution, from its

4 Michel Crozier, *The Bureaucratic Phenomenon* (The University of Chicago Press, Chicago, 1964) ; James G. March and Herbert A. Simon, *Organizations* (John Wiley, New York, 1958), ch. vii; Samuel P. Huntington, *The Common*

position as outside critic, may serve a useful function as a catalyst for innovation in the client agency. This "change agent" function may become important for many government agencies as developments at the frontiers of science and technology increasingly require rapid changes in objectives, plans, operating procedures, and organizational requirements. Government organization in the future will have to be unusually adaptive to keep pace with an accelerating rate of social change.

The nonprofit status of the advisory corporation like RAND also can offer a degree of impartiality and independence from the market that can be very useful to the government. In an era when the government must purchase expensive technologies from industry for various public purposes, it is important to have a source of advice apart from the industrial community on the costs and capabilities of advanced systems. Conflict-of-interest problems generally preclude a major role with respect to this kind of analytical service for the research institution operating under industrial auspices. Moreover, only the research organization completely divorced from the market seems able to gain access to proprietary industrial information needed for "state-of-the-art" surveys or other research purposes where knowledge of different firms' capabilities is required.

Finally, the advisory organization like RAND working in

Defense (Columbia University Press, New York, 1961), ch. v; Roland N. McKean, *Efficiency in Government Through Systems Analysis: With Emphasis on Water Resource Development* (John Wiley, New York, 1958), pp. 13-16, notes 21, 22 and 23; Morris Janowitz, *The Professional Soldier* (Free Press, Glencoe, 1962), ch. ii; James Q. Wilson, "Innovation in Organizations: Notes Toward a Theory," paper read at American Political Science Association, New York, September 1963; William H. Starbuck, "Organizational Growth and Development," in James G. March, ed., *Handbook of Organizations* (Rand McNally, Chicago, 1965), pp. 451-533; and work in progress by Anthony Downs, the RAND Corporation and the Real Estate Research Corporation.

302

the defense area can aid the civilian policy makers to achieve effective civilian control of the military (and conversely can educate the civilians on the military facts of life). To some extent this is a role shared with the entire community of civilian strategists and defense scholars, but primarily it belongs to the advisory organizations working in close contact with the government. The relationship between a vast military establishment and the civilian policy makers who direct that establishment has always posed vexing problems. Few societies have managed to secure effective civilian control of the military or to maintain stable democratic institutions over a period of years with a large military force in readiness. Nevertheless the United States has achieved under Secretary of Defense Robert S. McNamara perhaps a tighter civilian control of the military than has ever existed in modern American history. Part of the reason for McNamara's impressive feat has been his use of techniques of cost and budgetary analysis developed at RAND, and his selection for key posts of a number of talented individuals who had studied military problems professionally for some years as members of the RAND staff. The top civilian officials in the Pentagon have been able to group defense expenditures into functional categories more closely related to broad policy objectives and thus to achieve a strong policy control over the uniformed services through the budgetary process.[5] In a broad sense, moreover, the contribution of objective analyses has helped the civilian policy makers to achieve an understanding of the issues adequate to give critical examination to military tenets or "positions"—an understanding that many of their predecessors lacked or had to come by through instinct. Top defense

[5] William W. Kaufmann, *The McNamara Strategy* (Harper, New York, 1964), and Theodore H. White, "Revolution in the Pentagon," *Look*, April 23, 1963, pp. 31-49.

officials for the first time have been able to rely on civilians for informal advice on a wide range of questions concerning military affairs that previously were considered the professional military's exclusive domain. Without sources of advice available outside the military the task of civilian control would be considerably more difficult.

Ironically, this contribution to effective civilian control has posed perhaps the most troublesome problem that RAND now faces. New strains and stresses have arisen in the RAND–Air Force relationship. The Air Force, long RAND's major client, has viewed with some alarm the series of contract relationships that have developed between RAND and agencies at the Department of Defense level. The fear has arisen that RAND can no longer serve as the Air Force's confidential advisor since RAND, in effect, now works also for the boss and helps to control the military through the determination of broad strategic policy at the Department of Defense level. Many military officers, further, welcomed civilian advisors so long as the advisors served in purely advisory capacities and were clearly subordinate to the uniformed services. But deep-seated resentments against "defense intellectuals" have been stirred within the military now that some people who were formerly advisors have moved into policy-making positions above the military.[6] Also, in the past many potentially

---

[6] A good illustration of this point is the following statement by General Thomas D. White, retired Air Force Chief of Staff and for many years a strong supporter of RAND and the concept of civilian advisors aiding military planning: "In common with many other military men, active and retired, I am profoundly apprehensive of the pipe-smoking, tree-full-of-owls type of so-called professional defense intellectuals who have been brought into this Nation's capital. I don't believe a lot of these often over-confident, sometimes arrogant young professors, mathematicians and other theorists have sufficient worldliness of motivation to stand up to the kind of enemy we face . . ." General Thomas D. White, "Strategy and the Defense Intellectuals," *Saturday Evening Post*, May 4, 1963, pp. 10-12. For an analysis of the military "dispos-

troublesome disagreements with the client over strategic policy seldom reached serious proportions for RAND in part because the Air Force was in a position of undisputed leadership in the American military establishment and Air Force strategic doctrine generally shaped U.S. strategic doctrine. Now that the Air Force's position has become more vulnerable, it has become considerably sensitized to RAND's support or lack of support on broad strategic issues and crucial weapons acquisition choices.

We may pause here to note a general paradox of the contract advisory organization. The more "successful" the advisory organization—that is, in getting new points of view considered in policy formation—the harder it is to escape the pressures and constraints of the political process. In a sense, the successful advisory organization may contain the seeds of its own destruction. The group that was novel and imaginative, once it has become a serious influence, tends to be watched more closely and subjected to increasing client pressures which greatly complicate the research mission. The advisor's role as change agent is also affected since the various interests in the client agency tend to be more wary and circumspect in their dealings with the advisor. This sort of dilemma now seems to confront RAND. In order to retain its vitality and creativity as a research organization, RAND will have to make a conscientious effort to avoid being drawn into short-run policy disputes and lending prestige-type support to one current cause or another.

Does RAND have a future? Or is the independent advisory organization destined to be a passing phenomenon? Will there be a need for new generations of advisory groups to replace established organizations that have lost their vitality

sessed," see Daniel Bell's opening essay in Bell, ed., *The Radical Right*, rev. ed. (Doubleday Anchor, Garden City, N.Y., 1964), pp. 31-41.

or become "safe" appendages of the sponsoring agency?[7] Can one "plan" for the creation of new organizations that would duplicate the fortunate (and to a large extent accidental) circumstances of RAND's birth? Can advisory work for different clients with different interests be reconciled? In view of important changes in U.S. defense policy and administration, is there a more "normal" level in the defense hierarchy to which the advisory corporation should report than to a military service? Does "normal" in this context have any meaning other than the accidental sanctified by tradition?

The answers to these difficult questions, I think, might be essayed along the following lines. There are strong reasons for resisting further centralization in defense decision making. The defense establishment is simply too big and too complex to permit all important decisions to be made at one focal point. Subordinate elements in the defense hierarchy must continue to make complex choices; and in principle one can argue that advisory facilities should be available wherever they can contribute to more enlightened decisions on important problems. In this sense there is a rationale for a continuation of the RAND–Air Force relationship. Moreover, some sort of counsel of prudence and respect for functioning institutions may be relevant here. The question of the "logical" level for an advisory group to be formally attached to the government hierarchy perhaps has little meaning when custom and precedent have established a firm pattern. Some apparent oddities of departmentalization have proved workable once they have generated the momentum to be accepted as a normal part of the government landscape. About all that

---

[7] This was one reason Herman Kahn gave for leaving the RAND Corporation and setting up the Hudson Institute in 1961. He asserts that, whereas RAND functions as a client's "loyal opposition," he intends the Hudson Institute to serve as the "disloyal opposition." Personal interview.

one can say in the abstract is that the advisory group should normally report at a high enough level so that its work receives some serious attention. But it probably should not report at the highest policy levels because the difficulties of gaining access to information are increased and the political pressures are greatly multiplied. It is difficult to imagine, for example, that an independent advisory organization could operate very effectively at the White House level. The advisory group also probably could not achieve sufficient detachment from the political process to be able to perform a very useful research service at the congressional level. Similar difficulties might apply if RAND were to work exclusively for agencies at the Department of Defense level. The experience of the Institute for Defense Analyses (IDA) merits reflection in this context. It has been harder to make a success of IDA in part because this institution, closely tied to the high councils of policy where human purposes are in constant dispute, has inevitably had great difficulties in achieving a measure of detachment from the political process.

Despite difficulties, it would seem that RAND does have a future in approximately its present organizational form and without drastic changes in the current pattern of client support. There are still important mutual interests linking RAND and the Air Force. And with a growing professionalism in the advisory function, it should be possible for RAND to serve other government agencies at the same time, including agencies at the Department of Defense level.

The Air Force can continue to benefit from an association with RAND in a number of ways. There will be disagreements over questions of policy and strategic doctrine, but this is precisely where RAND has been very useful to the Air Force in the past. Heretical ideas generated at RAND in the long run have better equipped the Air Force to deal with important problems. Indeed, temporary strains and disagreements are an index of the healthiness of the advisory relationship. The advisory organization that functions in complete harmony with the client with no disputes on major issues is not doing its job effectively. Even the budgetary and cost-analysis techniques developed at RAND, which have given civilians more effective managerial control over the military services, have ultimately proved useful to the Air Force itself. The Air Force has begun to find that these tools give it more effective ways to manage its own internal affairs. Further, one should not overlook the fact that RAND has helped the Air Force to get its point of view heard and considered in the high policy councils and to adapt to the new style of decision making in the defense establishment. Perhaps more than either the Army or Navy, the Air Force has been able to adapt to the "analytic" style in defense decision making and to present its viewpoint in terms understandable to the top civilian officials.[8] RAND has played an important role here in providing a two-way communications link between the civilian controllers and the military service. Finally, whatever its ties at the higher policy levels, RAND can continue to provide a useful range of more or less purely "technical" advisory services for the Air Force's major technical commands.

For its part, RAND needs the stable sponsorship that the Air Force association has afforded over the years. The future would be uncertain without the stable support of a major client. Yet it is likely that RAND could benefit in the future from some further diversification in sponsorship to keep abreast of new research opportunities in areas related to its

[8] It is not accidental, for example, that the Navy felt disadvantaged and took pains to strengthen its advisory capabilities not after the Kennedy Administration took office. See Chapter II, p. 62.

work for the Air Force. To attract and retain a staff of high quality RAND must provide a range of stimulating research opportunities. The intimacy of the former special relationship with the Air Force will doubtless undergo changes. But with careful management it does not seem impossible for the independent advisory corporation to serve several clients successfully.

Quite possibly new cleavages in outlook and sources of conflict may develop within the organization, and a strain will be placed on management. The advisory organization, however, is *advisory* and needs no coordinated or consistent outlook. Its primary responsibilities are to maintain high professional standards in the research and to bring important policy alternatives to the decision maker's attention. The deleterious effects of conflict for an organization have often been greatly exaggerated.[9] Conflict may serve a healthy revitalizing function, particularly in an organization devoted to imaginative intellectual activity. A more complicated internal administration may bureaucratize RAND to some extent, but wise management should be able to protect the researcher's essential freedom and independence of operation. It should be possible also to develop some means for responsible private discussion with the Air Force or any other RAND client on a basis acceptable to the Secretary of Defense. Failing this, RAND might maintain a confidential advisory relationship with its different Pentagon clients by simply making it a matter of policy to decline certain tasks for agencies at the Department of Defense level which relate centrally to Air Force weapons systems and missions.

The future of the organization like RAND may parallel in some ways the evolution of the government corporation.

[9] See the works cited in note 3 of Chapter V.

At least one can discern some rough similarities between the path already traveled by the government corporation since the 1930's and the direction in which RAND seems to be moving. Both types of organizations arose in part out of a need to have a public function performed in a more flexible administrative framework than was easily available within one of the traditional executive departments. Both types of organizations initially enjoyed considerable autonomy in regard to personnel and staffing practices; the normal civil service pay scales did not apply and, consequently, employment opportunities tended to be more attractive than the civil service. As experience has been gained with each institutional device, however, there have been pressures to regularize its use. The desire to erase large salary differentials with the civil service and the need to establish some clear framework of public accountability for the corporation's operations were special points of concern in the case of the government corporations, and similar concerns now seem to be affecting RAND's development. The government corporations met with a series of efforts, culminating in the Government Corporation Control Act of 1945, to tie them more closely to the established governmental framework.[10] Some early observers were led to conclude that the government corporation had lost its distinctive characteristics and had ceased to exist as such.[11] Subsequent scholarship, however, has tempered these early judgments. A more accurate conclusion seems to be that the government corporations, although they have been integrated more closely into the executive hierarchy, retain

[10] Government Corporation Control Act, Public Law 248, 79th Cong., approved December 6, 1945.

[11] See, for example, C. Herman Pritchett, "The Paradox of the Government Corporation," *Public Administration Review* 1:381-389 (Summer 1941), and Pritchett, "The Government Corporation Control Act of 1945," *American Political Science Review* 40:495-509 (June 1946).

certain distinctive characteristics that continue to set them apart as a unique form of government organization.[12]

Similarly, the organization like RAND may face some additional controls but will perhaps still retain its identity as a novel organizational form. An example of a change that may affect all nonprofit corporations established under government auspices concerns the government's insistence on having first claim to all corporate assets in the event of dissolution.[13] Both the Research Analysis Corporation (RAC) and Aerospace have provisions to that effect in their charters or in their original contract documents. The System Development Corporation (SDC), which came into existence under RAND auspices, originally had a provision in its charters specifying reversion of assets to RAND upon dissolution for disposal in the public interest. But SDC in 1962, after the publication of the Bell Report, altered its charter to exclude RAND from any claim on assets in the event of dissolution. It is conceivable that RAND could be induced to alter its charter, and designate the government as receiver of corporate assets upon dissolution.[14] But so long as the crucial as-

[12] Harold Seidman, "The Government Corporation in the United States," *Public Administration*, Summer 1959 and Seidman, "The Theory of the Autonomous Government Corporation: A Critical Appraisal," in Dwight Waldo, ed., *Ideas and Issues in Public Administration* (McGraw-Hill, New York, 1953); and V. O. Key, Jr., "Government Corporations," in Fritz Morstein Marx, *Elements of Public Administration* (Prentice-Hall, New York, 1946). Seidman concludes that the government corporation has ". . . been admitted into the family of public institutions . . ." and ". . . has gained acceptance as a legitimate and appropriate form of Government organization." Seidman, "The Government Corporation in the United States," p. 105 and p. 114.

[13] The Bell Report stated as a general principle of equity that "where the Government has provided facilities, funds to obtain facilities, substantial working capital, or other resources to a contractor, it should, upon dissolution of the organization, be entitled to a first claim upon resources." Bell Report, in Holifield Subcommittee Hearings, part 1, p. 238.

[14] RAND's articles of incorporation specify that upon dissolution of the corporation all assets will devolve upon the Ford Foundation to be disposed of for such scientific and charitable purposes as it may direct. If the Ford

pects of independence are unaffected—that is, latitude in the program and publication areas—RAND can continue to perform a useful function as a novel gadfly critic for various government clients.

Government organization in general seems characterized by relatively long periods of routine and stagnation interspersed with short periods of change and experimentation.[15] The latter are often triggered by crises, disaster, or some other dramatic event that highlights an important need and jars loose the machinery of government sufficiently to permit major innovation in organization and procedure. In "normal" times, the tendency is to resist innovation and to bring novel experiments back into line with "traditional" administrative practices (unless the innovation has maintained momentum long enough to be accepted as part of the traditional pattern of organization). This line of reasoning has several possible implications for our discussion. In "normal" times, the circumstances may not be very favorable for the creation of the genuinely independent advisory organization. The independent advisory organization seems more likely to emerge in periods of flux when there is greater latitude to experiment with novel organizational forms. It follows also that the advisory organization will usually face a constant uphill struggle to preserve its independence and flexibility of operation. However, the record suggests that this struggle is eminently worthwhile and that there is a chance to persevere. Indeed, the very success of the advisory function seems to offer promise of an important new asset in the American governing system: the opportunity to "routinize" change to some

Foundation has ceased to exist, the articles of incorporation provide that the Superior Court of California will dispose of all RAND assets.

15 See the studies cited in note 4 of this chapter and John M. Gaus' treatment of the role of disaster in the "ecology" of government in his *Reflections on Public Administration* (University of Alabama Press, Birmingham, 1947).

extent. Through the constant stimulus of a detached yet friendly critic, government agencies may be in a position to anticipate some major problems and to make changes in policies and procedures before disaster has dramatized the need for innovation.

Can the RAND-type advisory institution be usefully employed in nondefense areas? This question seems especially pertinent in view of a persistent belief in some quarters that the defense advisory organizations have merely acted as a subtle means to advance client interests. In particular, the view is often expressed that RAND (and similar advisory organizations) have served to block efforts toward peace by accustoming people to think in warlike categories.[16] Under this view, the advisory organization is thought to possess a "mind" which inevitably reflects client interests. The following is a fairly typical statement of this point of view: "The RAND Corporation is financed mainly by the United States Air Force, so that its studies must be accepted with the same kind of reserve that, shall we say, we might greet a study of the Reformation by Jesuits based on unpublished and secret documents in the Vatican; there is the same combination of honesty in the value system and bias in the commitment."[17]

In actuality, the fears of RAND manipulating attitudes in an unhealthy direction seem greatly exaggerated. There is no more a RAND "mind" than there is an academic, scientific, or military "mind." Although RAND studies naturally tend to center in subject matter areas of interest to its clients, RAND internally is far from monolithic. RAND studies are the products of individuals, and reflect the variety of outlooks

---

[16] For example, Saul Friedman, "The RAND Corporation and Our Policy Makers," *Atlantic Monthly* 212:61-68 (September 1963).

[17] Kenneth E. Boulding, *Conflict and Defense* (Harper & Row, New York, 1962), p. 332.

characteristic of a group of diverse talents. The fact that RAND has done studies for such nonmilitary clients as the Arms Control and Disarmament Agency, the Agency for International Development, and the National Aeronautics and Space Administration lends substance to the view that there is no rigid RAND "mind" preoccupied with finding ways to continue the cold war. It is therefore difficult to credit the notion that RAND, as it is presently staffed and operated, acts as a subtle instrument to subvert the thinking of influential citizens and propagandize for a certain policy stance on cold war issues. If RAND were to cease being a research organization and instead became an "action" organization, a very different situation would be presented and some serious issues would be raised. Under such circumstances, it would certainly be proper to question whether tax moneys should support a private organization in carrying on publicity campaigns and propaganda activities in an effort to shape public attitudes and influence legislation.[18] However, this is distinctly not the situation now. RAND's nongovernment contacts are primarily with the intellectual community, not with the public directly, and the main effort is to stimulate scholarly research. Moreover, the pluralism of the advisory system, in

---

[18] It will also be recalled that a nonprofit organization would jeopardize its tax-exempt status if a substantial part of its activities were "carrying on propaganda, or otherwise attempting, to influence legislation." A small pacifist nonprofit organization, the Fellowship of Reconciliation, recently lost its tax-exempt status for this reason. The U.S. Internal Revenue Service formally revoked the organization's tax-exempt status on January 10, 1963, on the grounds that the Fellowship of Reconciliation was an "action" organization whose main objectives could be attained only by legislation or defeat of proposed legislation. Earlier, on September 26, 1962, New York State Supreme Court Justice James W. Bailey in Cold Springs, New York upheld the Fellowship's right to exemption from local real estate taxes against a claim by the Village of Upper Nyack and the Town of Clarkstown. See "United States Revokes Tax Exemption Held by Pacifist Group Since '26," *New York Times*, May 13, 1963, p. 1, col. 7.

which RAND is only one institution among many with access to persons in authority, helps assure that no one group will monopolize the attention of policy makers.

A more subtle variant of the argument depicts the institution like RAND as subverting the nation's intellectual skills into narrow short-run tasks at the expense of detached scholarship and the intellectual's long-range social critic function. There is some justice to this argument. A healthy society must surely provide ample opportunity for pure scholarship, and protect the intellectual's right and incentive to criticize prevailing assumptions and social values, but eliminating the RAND-type institutions would not eliminate the ever-increasing public demands on society's intellectual skills. The institution like RAND, far from subverting intellectual skills to a dangerous extent, serves an important function as a kind of institutional buffer between the academic community and growing public demands on intellectual elites.

Furthermore, the belief that there is an obvious "war" as opposed to a "peace" position on national-security issues begs an important intellectual question that deserves to be brought to light. On most complex defense issues, it is not at all apparent what course of action will reduce tensions, stabilize the international environment, and minimize the likelihood of conflict. Certainly there is little basis for making *a priori* judgments about the efficacy of a given set of policy initiatives. Broad defense issues involve an extraordinarily complex mixture of technological considerations, conflicting objectives, and assessments of uncertainties in which experience and intuition alone are poor guides to decision. Research and analysis have been able to play an important role by helping the policy maker to define the crucial dimensions of complex problems. There have doubtless been occasions when

315

analytic studies have made unrealistic assumptions, drawn false or misleading inferences from empirical data, and failed to take into consideration vital aspects of a broad problem. But this is not an argument against the advisory function in principle: this is an argument for more and better analysis and an increased professionalism in the advisory ranks.

The "analytic" style in defense decision making is probably an inescapable concomitant of the growing complexity of defense issues. The infusion of an accelerating technology into the military arena helped make many of the choices facing the defense agencies almost incomprehensible without the aid of analytical expertise. This provides a more persuasive explanation than conspiracy theories of why the advisory organizations arose first in the defense area. The role played by science and technology in World War II was apparent and the military needed scientific expertise of various kinds to supplement the skills of staff officers. The Air Force was the first service to experiment with the novel advisory corporation because it was less bound by organizational precedent and its mission was most powerfully influenced by science and technology. It was perhaps fortunate that RAND began as a "technical" advisor and operations researcher in view of the National Science Foundation's difficulties with the social science issue in the early period.[19]

[19] For a discussion of these difficulties, see Don K. Price, *Government and Science* (New York University Press, New York, 1954), ch. ii, "Freedom or Responsibility?" A similar problem arose with respect to the Bureau of Agricultural Economics. The Bureau found its budget severely cut by the Appropriations Committee of the House of Representatives because, in the eyes of some Congressmen, the Bureau had gone beyond the business of scientific research and had undertaken in addition to help the Secretary of Agriculture plan his policies and program. See Charles M. Hardin, "The Bureau of Agricultural Economics under Fire: A Study in Valuation Conflicts," *Journal of Farm Economics* 28:635-660 (August 1946).

If RAND had been a policy-research organization from the start, with numerous social skills on the staff, one may wonder whether it could have escaped congressional scrutiny of some kind in the 1945-1950 period. At any rate, the defense area was the one area of public policy where Congress and the public were willing to grant the expert wide latitude and to relax somewhat the normal channels of political accountability. Use of the contract advisory organization gradually spread to other defense agencies. And finally, to greater or lesser degrees, the defense advisory organizations evolved a technical-policy advisory function as it became apparent that "technology" and "policy" could not be sharply separated. Hence the "analytic" style in defense policy formation must be considered a product of the scientific revolution of the twentieth century.

As developments in science and technology increasingly affect other areas of public policy, there can only be a growing need for the kind of advisory service that RAND has provided various defense clients. In general, it would seem highly desirable for similar advisory capabilities to be built up by nondefense agencies. Indeed, one may put special stress on the need for advisory services in areas where defense concerns shade into the concerns of civilian agencies. The way to assure that important alternatives are not neglected in the processes of policy formation is to develop the broadest possible advisory base. The important objective should not be to degrade one set of intellectual capabilities but to concentrate on ways to build up such capabilities in the civilian agencies.

One sees such a trend already in evidence. Interestingly, the State Department, an agency that parallels the Defense Department in needing long-range thinking to perform its mission effectively, until recently had almost no tradition of support for outside policy research activities. A strong traditional "humanist" bias against the social sciences, and perhaps the sciences in general, probably contributed to the State Department's failure to develop any significant program of research support. In 1961, however, the affiliated Agency for International Development (AID) embarked on a substantial program of supporting analytical research of various kinds.[20] The National Aeronautics and Space Administration (NASA) has developed a "think" corporate entity—Bellcomm, Inc.—whose relationship with NASA bears a striking affinity to RAND's relationship with the Air Force.[21] The Arms Control and Disarmament Agency (ACDA) has a con-

[20] In the Foreign Assistance Act of 1961, the Congress, at the request of the executive branch, enacted a provision which for the first time authorized research to make more effective use of foreign aid funds. Title V, Section 241 of that act provided: "The President is authorized to use funds made available for this part to carry out programs of research into, and evaluation of, the process of economic development in less developed friendly countries and areas, into the factors affecting the relative success and costs of development activities, and into the means, techniques, and such other aspects of developmental assistance as he may determine, in order to render such assistance of increasing value and benefit."

For a description and criticism of the resulting AID program of contract research, see *Agency for International Development Contract Operations*, Twenty-Third Report by the Committee on Government Operations, 87th Cong., 2nd Sess., part 1, Union Calendar No. 1017, September 19, 1962, and Twenty-Sixth Report, part 2, Union Calendar No. 1037, September 26, 1962, and *Agency for International Development Contract Operations*, Hearings before a Subcommittee on Government Operations, U.S. House of Representatives, 87th Cong., 2nd Sess., parts 1 and 2, August-September 1962.

[21] See *Systems Development and Management*, Hearings before a Subcommittee of the Committee on Government Operations, U.S. House of Representatives, 87th Cong., 2nd Sess., part 5, August 1962, pp. 1751-1776 (the Holifield Subcommittee Hearings). The following exchange took place (p. 1758) between Mr. Herbert Roback, committee staff administrator, and Mr. John Hornbeck, President of Bellcomm, Inc., respecting the Bellcomm resemblance to RAND:

"Mr. Roback: . . . would this be a fair analogy, that this [Bellcomm, Inc.] is a kind of profit RAND Corporation, a profit RAND for NASA; would that be approaching the type of services that are done, say, by RAND for the Air Force?

Mr. Hornbeck: There are certainly elements in common, strong elements in common."

318

tract research program totaling $11,000,000 for analytical studies of arms control and disarmament issues.[22] Even in such an area as water resource policy, traditionally the domain of "pork barrel" domestic politics, the technical complexity of the issues has forced government agencies to contract for analytical assistance in policy formation.[23]

It is interesting to speculate about the success of possible applications of systems-analysis techniques to the problems of the civilian agencies. There is some reason to believe that such capabilities are badly needed in the domestic arena and that they could contribute impressively to the operations of many nondefense agencies. Salient aspects of domestic problems will involve the complex interaction of technical and political considerations, and systematic research employing diverse professional skills may be expected to help clarify difficult choices at various policy levels. However, there are apt to be some formidable difficulties in the case of the domestic agencies which may require modification in the kinds of analysis that can be fruitfully applied, the balance of professional skills on the research team, and the method of operation of the advisory unit. The defense agencies, after all, have fairly well-defined missions that seem to lend them-

[22] See *Congressional Record*, Senate, June 13, 1963, pp. 10236-10241.

[23] The Army Corps of Engineers in the late 1950's contracted with the Harvard Water Resources Program of the Graduate School of Public Administration, Harvard University, for analytic studies of water resource development problems. For a brief description of the program, see *Public Works Appropriation for 1962*, Hearings before a subcommittee of the Committee on Appropriations, U.S. House of Representatives, 87th Cong., 1st Sess., part 2, pp. 2290-2. A substantial study of questions of methodology and systems design in this area of public policy resulted. See Arthur Maass, Stephen A. Marglin, et al., *Design of Water Resource Systems: New Techniques for Relating Economic Objectives, Engineering Analysis and Government Planning* (Harvard University Press, Cambridge, Mass., 1962). An earlier RAND study by Roland N. McKean, *Efficiency in Government Through Systems Analysis: With Emphasis on Water Resource Development* (John Wiley, New York, 1958), had helped to stimulate interest in this area.

selves well to systematic analysis. It is harder to know what might be a preferred "system," or how to assess the cost and effectiveness of various actions, when one considers the amorphous problems of, say, the Department of Health, Education, and Welfare or the Office of Economic Opportunity. Possibly social science skills would have to be more prominently represented on the staff of a domestic "RAND" and some stable pattern of interaction evolved with Congress as well as with the executive sponsor. We may see new institutional forms that differ from RAND, and changes in government organization may bring some part of the analytic function into the framework of the traditional executive agencies. But it seems clear at least that we are moving into a period in which science and technology will have a growing impact on both the structure of government and the content of public policy.

Science and technology, by introducing so many complexities into public policy, have destroyed the comfortable nineteenth-century notion that public issues could really be determined by the clash of political ideologies.[24] Important implications for the future of our governing system inevitably follow. The "intellectual" content of policy debate will likely be greatly elevated. Traditional elite groups like the lawyers will have to share influence with newer scientific and intellectual elites. Enormous demands will be made on private institutions and skills in the pursuit of public objectives. Expertise will narrow and focus many of the choices open to accountable officials, and relegate some cherished beliefs to the penumbra of serious discussion. This is not to suggest that expertise can ever replace politics and the political process. Value

[24] See the discussion of this point in Don K. Price, "The Scientific Establishment," *Proceedings of the American Philosophical Society* 106:239-340 (June 1962).

conflicts remain the essence of politics. Even in a society of abundance, there will be disputes about the division of life's good things (and disagreements about what the good things are) which provide the stuff of the political process. What the growing complexity of public affairs implies, rather, is that fruitful policy debate will tend to be marked less and less by polemic and first principle. Instead, discussion will center more in serious analysis of alternative means to achieve common ends and, where the ends themselves are unclear or in dispute, in systematic consideration of the costs and consequences of pursuing different ends. Used properly, the advisory institution like RAND can contribute to sensible policy decisions and can help to maintain the dynamism of America's pluralist governing system.

In the net, the device of the RAND-type advisory corporation seems to reflect both the strengths and weaknesses of American pluralism. The presence of a number of advisory institutions like RAND helps to assure decision makers of a broad base of scientific advice and to guard against the dangers of a closed system with a narrow technocracy cut off from the healthy effect of outside criticism. There is very little danger that anything like a monolithic statism or a vast bureaucracy on Weberian lines will emerge, given a system which decentralizes expertise and influence throughout many different institutions in society. The real dangers of this system may not be what is commonly supposed. The chief danger may well be that familiar problem of American politics: keeping organizational pluralism within some sort of bounds so that a framework for coherent, unified, and sustained national policies can be maintained. If there is a growing need for professionalism in the advisory ranks, there is surely an even greater need for professionalism among our generalist policy makers (especially at the top career and lower political executive levels). For it is these men who must understand the potentialities and limitations of expert advice as an aid to policy formation. These individuals must distill numerous points of view, expert and nonexpert alike, into a coherent policy and action program. A system of extensive contract-advisory services keeps new ideas flowing to the points of decision, and guarantees that some important problems will be studied in depth. But it does not guarantee the government's ability to maintain a clear central direction in policy and to keep a steady and controlling hand on its advisory apparatus. This suggests that it will be important to enhance the attractiveness of the government as an employer so that a fair share of society's scarce technical and intellectual talents will seek a public career. It also suggests that we will have to make discriminating use of the various institutional arrangements that exist in the broad partnership between science and government. Avoiding the evils of a centralized bureaucracy may only mean that we encounter the subtle evils of a network of private bureaucracies. The challenge is to make use of institutional specialization in the policy process without sacrificing the integrity and discipline and direction of the total effort. It is clear at least that we must avoid such a diffusion of expertise and influence that consistent, effective, and responsible public action becomes impossible. The formal government must remain the political center of gravity. Or else, in making our governing system complex, we will have made it unmanageable.

INDEX

Acheson, Dean, 23n34
Ackoff, Russell L., 280n56
Administration, *see* Organization, internal
Advanced Research Projects Agency (ARPA), 126, 270-271
Advertising, institutional, 100
Advice: role in policy formation, 11-12, 27, 195-199; limitations of, 27-28, 321-322
Advisory corporations, nonprofit, 1-6, 19, 31-33, 69, 105-107; critics of, 14-18 passim, 140-142 passim; nonprofit *vs.* profit, 249-256. *See also* Bell Report; Civilian, agencies and advisory service; Conflict-of-interest; Congress, and "nonprofits"; Independence, questions of; Industry, and "nonprofits"; Politics, dangers of; Role, present and future; Tax exemption privileges
Aerospace, 4, 17, 20, 69, 101-102, 118n38, 250-251, 287-288n63, 311
Agency for International Development (AID), 128, 164, 314, 318
Air Defense Direction Center, research at, 115
Air Force, 49n22; advisory corporations of, 3-4; Air Force Council, 221n19; officers on assignment at RAND, 168, 169; policy statement, 21 July 1948, 76-85; policy statement on nonprofits, 134-135; relations with RAND, 52, 62, 76-85 passim, 91-92, 133-135, 304-308 passim; Space Technology Laboratory, 102n16;
    studies: Air Defense Study, 104-

105n21; B-29 Special Bombardment Project, 35-38, 42, 64; Offensive Bomber Study, 104-105n21. *See also* Strategic Bases Study
    See also Directorate of Development planning; Diversification in sponsorship; Douglas Aircraft Company; Military, new role of civilians in
Air Force Systems Command, 102, 121
Air Research and Development Command, 121
American Meterological Society, 109
American Telephone and Telegraph, 252
Analysis, tools of, 110-111; changing nature of on broad strategic issues, 103-105, 198, 210-211; differences in, 8-10
Analytic Services, Inc. (ANSER), 3-4, 75, 76, 90, 106, 119-125, 246, 267n35, 288
"Analytic" style in defense decision making, 28, 316-317
Anderson, Theodore, 130n43
Andrews, F. Emerson, 68n3, 185n30, 188n38, 189n39, 189n40
Ansoff, H. Igor, 209
Anti-submarine Warfare Operations Group (ASWORG), 2, 34
Appleby, Paul, 217n18
Arms Control and Disarmament Agency (ACDA), 314, 318-319
Army, 2-3, 133, 271-273, 308
Army Corps of Engineers, 319n23
Arnold, General H. H., 36, 38, 40-43 passim, 45, 48, 49n22
Arnow, Kathryn S., 300n3

HARVARD POLITICAL STUDIES

* Out of print

John Fairfield Sly. *Town Government in Massachusetts (1620–1930)*. 1930.*

Hugh Langdon Elsbree. *Interstate Transmission of Electric Power*. 1931.*

Benjamin Fletcher Wright, Jr. *American Interpretations of Natural Law*. 1931.*

Payson S. Wild, Jr. *Sanctions and Treaty Enforcement*. 1934.*

William P. Maddox. *Foreign Relations in British Labour Politics*. 1934.*

George C. S. Benson. *Administration of the Civil Service in Massachusetts*. 1935.*

Merle Fainsod. *International Socialism and the World War*. 1935.*

John Day Larkin. *The President's Control of the Tariff*. 1936.

E. Pendleton Herring. *Federal Commissioners*. 1936.*

John Thurston. *Government Proprietary Corporations in the English-Speaking Countries*. 1937.*

Mario Einaudi. *The Physiocratic Doctrine of Judicial Control*. 1938.

Frederick Mundell Watkins. *The Failure of Constitutional Emergency Powers under the German Republic*. 1939.*

G. Griffith Johnson, Jr., *The Treasury and Monetary Policy, 1933–1938*. 1939.*

Arnold Brecht and Comstock Glaser. *The Art and Technique of Administration in German Ministries*. 1940.*

Oliver Garceau. *The Political Life of the American Medical Association*. 1941.*

Ralph F. Bischoff. *Nazi Conquest through German Culture*. 1942.*

Charles R. Cherington. *The Regulation of Railroad Abandonments*. 1948.

Samuel H. Beer. *The City of Reason*. 1949.*

Herman Miles Somers. *Presidential Agency: The Office of War Mobilization and Reconversion*. 1950.*

Adam B. Ulam. *Philosophical Foundations of English Socialism*. 1951.*

Morton Robert Godine. *The Labor Problem in the Public Service*. 1951.*

Arthur Maass. *Muddy Waters: The Army Engineers and the Nation's Rivers.* 1951.

Robert Green McCloskey. *American Conservatism in the Age of Enterprise.* 1951.*

Inis L. Claude, Jr. *National Minorities: An International Problem.* 1955.*

Joseph Cornwall Palamountain, Jr. *The Politics of Distribution.* 1955.*

Herbert J. Spiro. *The Politics of German Codetermination.* 1958.*

Harry Eckstein. *The English Health Service.* 1958.

Richard F. Fenno, Jr. *The President's Cabinet.* 1959.

Nadav Safran. *Egypt in Search of Political Community.* 1961.

Paul E. Sigmund. *Nicholas of Cusa and Medieval Political Thought.* 1963.

Sanford A. Lakoff. *Equality in Political Philosophy.* 1964.

Charles T. Goodsell. *Administration of a Revolution.* 1965.

Martha Derthick. *The National Guard in Politics.* 1965.

Bruce L. R. Smith. *The RAND Corporation: Case Study of a Nonprofit Advisory Corporation.* 1966.